Final Cut Pro X for iMovie and Final Cut Express Users
Making the Creative Leap

Tom Wolsky

ELSEVIER

AMSTERDAM • BOSTON • HEIDELBERG • LONDON
NEW YORK • OXFORD • PARIS • SAN DIEGO
SAN FRANCISCO • SINGAPORE • SYDNEY • TOKYO

Focal Press is an imprint of Elsevier

Focal
Press

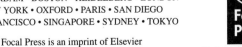

Focal Press is an imprint of Elsevier
225 Wyman Street, Waltham, MA 02451
The Boulevard, Langford Lane, Kidlington, Oxford, OX5 1GB, UK

Notices

Knowledge and best practice in this field are constantly changing. As new research and experience broaden our understanding, changes in research methods, professional practices, or medical treatment may become necessary.

Practitioners and researchers must always rely on their own experience and knowledge in evaluating and using any information, methods, compounds, or experiments described herein. In using such information or methods they should be mindful of their own safety and the safety of others, including parties for whom they have a professional responsibility.

To the fullest extent of the law, neither the Publisher nor the authors, contributors, or editors, assume any liability for any injury and/or damage to persons or property as a matter of products liability, negligence or otherwise, or from any use or operation of any methods, products, instructions, or ideas contained in the material herein.

Library of Congress Cataloging-in-Publication Data
Wolsky, Tom.
 Final Cut Pro X for iMovie and Final Cut Express users : making the creative leap / Tom Wolsky.
   pages cm
 ISBN 978-0-240-82366-9 (pbk.)
 1. Video tapes – Editing – Data processing. 2. Digital video – Editing – Data processing.
 3. Final cut (Electronic resource) 4. iMovie. I. Title. II. Title: Final Cut Pro Ten for iMovie and
   Final Cut Express users.
 TR899.W697 2012
 777'.55—dc23

                                2011047376

British Library Cataloguing-in-Publication Data
A catalogue record for this book is available from the British Library.

For information on all Focal Press publications
visit our website at *www.elsevierdirect.com*

12 13 14 15   5 4 3 2 1

Printed in the United States of America

Typeset by: diacriTech, Chennai, India

For B.T.
With Endless Love and Gratitude

Contents

Acknowledgments

First, as always, my gratitude to all the people at Focal Press who make this book-writing process relatively painless, particularly Dennis McGonagle, acquisitions editor, for his thoughtful advice and guidance, and for bringing this project to life. My thanks also to Carlin Reagan, senior editorial project manager, and Lauren Mattos, editorial project manager, for their help and ready answers to my questions and for shepherding the project through the production process. Special thanks to project manager Sarah Binns for helping me through the final proofing process. Many thanks are due to Chuck Hutchinson for his stylish copyediting and to Joanne Blank for her work on the wonderful cover. My thanks to diacriTech for proofreading, typesetting, and carefully indexing the book. Any errors in text or substance remain mine alone.

So many people helped make this book possible and deserve thanks: special thanks to Loren Miller for taking on the task of doing the technical edit for this book, on an application he does not like to use; Sidney Kramer for his expert contract advice; and Steve Martin of Ripple Training, the trainers' trainer, who taught us the technique of using sparse images for instructional media.

My thanks go to Rich Lipner of Finca Dos Jefes and Café de la Luna for his gracious cooperation for allowing us to shoot his coffee tour, with special thanks to Lazario and Hermione for their assistance. I have to thank Travis Schlafmann for letting me use the great snowboarding and skiing video that accompanies the book and appears on the cover. Thanks as always to our friends in Damine, Japan. A great many thanks are due to my partner, B. T. Corwin, for her insights, her proofing, her endless encouragement, her engineering technical support, and her patience with me. Without her, none of this would have been possible. Any errors, mistakes, or omissions are neither BT's nor Loren's fault, but solely mine as I've looked at these words and thought about them long enough.

Introduction

WHAT IS EDITING?

The first movies were single, static shots of everyday events. The Lumière brothers' screening in Paris of a train pulling into the La Ciotat train station caused a sensation. Shot in black and white, and silent, it nevertheless conveyed a gripping reality for the audience. People leaped from their seats to avoid the approaching steam locomotive. The brothers followed this with a staged comic scene. Georges Méliès expanded on this by staging complex tableaux that told a story. It wasn't until Edwin H. Porter and D. W. Griffith in the United States discovered the process of editing one shot next to another that movies were really born. Porter also invented the close-up, which was used to emphasize climactic moments. Wide shots were used to establish location and context. Griffith introduced such innovations as the flashback, the first real use of film to manipulate time. Parallel action was introduced, and other story devices were born, but the real discovery was that the shot was the fundamental building block of film and that the film is built one shot at a time, one after the other. It soon became apparent that the impact of storytelling lies in the order of the shots.

Films and videos are made in the moments when one shot changes into another, when one image is replaced by the next, when one point of view becomes someone else's point of view. Without the image changing, all you have are moving pictures. The idea of changing from one angle to another or from one scene to another quickly leads to the concept of juxtaposing one idea against another. The story is not in each shot but in the consecutive relationship of the shots. The story isn't told in the frames themselves so much as in the moment of the edit, the moment when one shot, one image, one idea, is replaced with another. The edit happens not on a frame but between the frames, in the interstices between the frames of the shots that are assembled.

Editing is a poor word for the film or video production process that takes place after the images are recorded or created. It is not only the art and process of selecting and trimming the material and arranging it in a specific order but is very much an extension of script writing, whether the script was written before the project was recorded or constructed after the material was gathered and screened. The arrangement and timing of the scenes, and then the selection, timing, and arrangement of the picture elements within each scene, are most analogous to what a writer does. The only difference is that the writer crafts his script from a known language and his raw imagination, whereas the editor crafts her film from the images, the catalogue of words, as it were—the dictionary of a new language—that has been assembled during the production. The editor's assembly creates a text that, although a new, never before seen or heard language, is based on a grammatical tradition—one that goes back to Porter and Griffith—that audiences have come to accept as a means for conveying information or telling a story.

Unlike the written language, a novel, or an essay that can be started and stopped at the reader's whim, video or film production is based on the concept of time, usually linear time of a fixed length. Nowadays, of course, many forms of video delivery—the Web, computer or portable player, or DVD player—can be stopped and started according to the viewer's desires. Nonetheless, most productions are designed to be viewed in a single sitting for a specified duration.

Film and video are designed to accommodate the temporal rigidity of the theater but with the spatial fluidity and freedom of a novel. Whether it is 10 minutes, 30 minutes, one hour, two hours, or more, the film is seen as a single, continuous event of fixed duration. On the other hand, time within the film is infinitely malleable. Events can happen quickly: We fly from one side of the world to another, from one era to a different century, in the blink of an eye. Or every detail and every angle can be slowed down to add up to a far greater amount than the true expanse of time, or the images can be seen again and again.

Because film and video production are based on the notion of time, the process of editing—controlling time and space within the story—is of paramount importance. This process of editing, of manipulating time, does not begin after the film is shot but as soon as the idea is conceived. From the time you are thinking of your production as a series of shots or scenes and as soon as the writer puts down words on paper or the computer, the movie is being edited, and the material is being ordered and arranged, juxtaposing one element against another, one idea with another.

This process of writing with pictures that we call editing has three components:

1. *Selection*, choosing the words, deciding which shot to use
2. *Arrangement*, the grammar of our writing, determining where that shot should be placed in relation to other shots, the order in which the shots will appear
3. *Timing*, the rhythm and pace of our assembled material, deciding how long each shot should be on the screen

Timing is dictated by rhythm—sometimes by an internal rhythm the visuals present, sometimes by a musical track, and often by the rhythm of spoken language. All language, whether dialog or narration, has a rhythm, a cadence or pattern, that is dictated by the words and based on grammar. Grammar marks language with punctuation: Commas indicate short pauses, semicolons are slightly longer pauses, and periods mark the end of a statement. The new sentence begins a new idea, a new thought, and it is natural that as the new thought begins, a new image is introduced to illustrate that idea. The shot comes not at the end of the sentence, not in the pause, but at the beginning of the new thought. This is the natural place to cut, and this rhythm of language often drives the rhythm of film and video. Editing creates the visual and aural juxtaposition between shots. That's what this book is about: how to put together those pieces of picture and sound.

This book is not intended as a complete reference manual for all aspects of Final Cut Pro. For that you would really need to consult the manual in the Help files that accompanies the application. Nor does this book cover any ancillary applications that may be used with FCP, such as those bundled with earlier versions of Final Cut Pro, or now sold separately, such as Compressor or Motion. Rather this book is intended as a course to give you a good grounding and understanding for practical use of the application. If there is something specific you'd like to learn how to do, I suggest looking first in the index to see if it's covered in the book.

There are many different ways of doing things in FCP, different ways to edit and to trim, but there are clearly more efficient ways. There are many ways to do things in Final Cut because no one way is correct in every situation. Working cleanly and efficiently will improve your editing by allowing you to edit more smoothly and in a steady rhythm, concentrating on content rather than on the mechanics of the application.

The book is organized as a series of tutorials and lessons that I hope have been written logically to guide the reader from one topic to a more advanced topic. The nature of your work with Final Cut Pro, however, may require that you have the information in Lesson 10, for example, right away. You can read that lesson by itself. There may, however, be elements in Lesson 10 that presuppose that you know something about using the Viewer in conjunction with the Inspector.

WHY DO I GET TO WRITE THIS BOOK?

I have been working in film and video production for longer than I like to admit (okay, more than 40 years). I worked at ABC News for many years as an operations manager and producer, first in London and then in New York. I went on to teach video production at a small high school in rural northern California. I wrote the first step-by-step tutorial book with project files and media for Final Cut Pro in 2001. Since then I have written many FCP and FCE books as well as made training DVDs for Final Cut Pro, Final Cut Studio, Final Cut Express, iLife, and the Mac OS. I also have written curriculum for Apple's Video Journalism program and taught training sessions for them, and during the summer, I have had the pleasure of teaching Final Cut at the Digital Media Academy on the beautiful Stanford University campus.

WHO SHOULD READ THIS BOOK?

This book is intended for all FCP users, especially those new to the application and those moving up from consumer products like iMovie and Final Cut Express. Long-time FCP users will benefit greatly from this book as well because this is a completely new application. Apple, I think, sees this new application at consumer pricing as a paradigm shift for the professional market. Apple has always been about innovation and moving the playing field, not making the game better, but changing the game. Whether this will work for professionals remains to be seen, but releasing the product at consumer pricing means it will have great appeal and great penetration, which Apple believes will become the norm in the professional market as the first Final Cut Pro application did.

While Final Cut Pro is directed to professional video and film production market, its price point and ease of use make it appealing to consumers, prosumers, and professionals, really any users making video or film from any format for any market should find this application fits their needs. Its affordability will particularly benefit the prosumer market, event producers, and even small companies with video production requirements. I think it's a truly great video application for education—fully featured, able to go beyond the limitations that frustrate many students who use iMovie, at a price that makes it accessible even to school budgets in these penny-pinching times. Final Cut Pro is a complex application. It requires learning your way around the interface, its tools, and its enormous capabilities.

WHERE'S THE MEDIA?

All of my previous books have had an accompanying DVD. In these days of cloud computing, this is no longer felt to be necessary. All the project files and media are now stored on the servers. To download the media, go to http://www.fcpxbook.com. On the website there are a number of zipped files. These contain sparse images. These are expandable disk images, so the initial size is not so large, but each image can be expanded up to 10GB. These ZIP files are all fairly large and will not download instantly. If you want, you can download them all at once or as you need them. With the exception of the file called *NONAME*, they are numbered sequentially as you use them in the book. I suggest keeping the original ZIP files after you download them so that you can go back to the original media whenever you need to.

The first lesson does not need any media, but most will, so the sooner you can start downloading, the better. You may want to substitute your own material—clips you want to work with or are more familiar with, but I suggest starting with the supplied materials.

I hope you find this book useful, informative, and fun. I think it's a great way to learn this kind of application.

Getting Started with Final Cut Pro X

Welcome to Final Cut Pro X (pronounced *ten*). The application is listed as version 10, but to be honest it really is the first version of a brand new application for video production. This is unlike any previous versions of Final Cut Pro or Express. On the surface it is more like iMovie than it is similar to traditional video editing applications like legacy Final Cut Pro, Adobe Premiere, or Avid Media Composer. It is entirely new, a completely new concept for professional video editing built on an entirely new foundation, using the most modern technologies available in the operating system, 64-bit, OpenCL, Grand Central Dispatch, Core Video, and others. You don't have to understand what these do except to know that they allow the application to use all the available power in your computer, as much RAM as you have, as many processors as are in your machine, as well as utilizing your graphics processor, all to speed up, simplify, and make the application more capable. The application is called X because it is based around the capabilities of the Mac OS X operating system, unlike previous versions that were first introduced and used on Mac OS 8.5, and have since been strapped onto the newer and more sophisticated operating systems.

iMOVIE AND FINAL CUT EXPRESS

Final Cut Pro X, which we will simply call Final Cut Pro or FCP, is unique in video editing applications. It is the first professional application capable of editing everything from consumer video to high-resolution, film-size formats, while still being sold at a consumer price. Until now those who were interested in producing videos could either use iMovie, which comes free, pre-installed on all Macs, or step up to an application like Final Cut Express, or even one of the full-featured professional applications but at substantial cost.

Many users' first foray into video production is with iMovie. Serious hobbyists and professionals starting to work with the application often became quickly frustrated by that application's limitations and quality issues and had to step up to legacy Final Cut Pro's younger brother, Final Cut Express. At that time Final Cut Pro was available only as part of a suite of applications called Final Cut Studio, most of which the majority of consumers and many professionals simply did not need, nor were they willing to pay for. Final Cut Express was designed to occupy the space between iMovie and Final Cut Pro at a consumer affordable price. For many, the move to Final Cut Express, though affordable, was fraught with dangers. The application was hugely different from iMovie, a complete paradigm shift to a professional interface with two monitors and multiple tracks as they were called in FCE.

With the new Final Cut Pro, this has changed. The new application will not be as strange and intimidating as its predecessors to iMovie users. In fact it is very much based on the iMovie interface. Much of what was familiar in iMovie will be there in FCP. As you start to work with this application, you will find much that is recognizable: the layout of the interface, the terminology, and even many of the editing functions and tools. In fact it will be legacy Final Cut Pro and Final Cut Express users who will find the application strange and foreign. The entire interface and especially a completely new glossary of terms will stand between them and this new application. In the beginning everything that will be familiar to iMovie users will be new and strange to Final Cut users. The further we go into our exploration of this application, the more the techniques and tools will become familiar to Final Cut users and the more strange and complex they will be for iMovie users. As you go through the lessons in the book, I will try to point out the differences and similarities in the interface and in the workflow for both users.

The first few lessons will be familiar ground to iMovie users and strange to Final Cut users, covering whole new ways of dealing with media and how it's stored and cataloged and organized and edited into a very new Timeline. As the lessons progress, the interface and capabilities of FCP will become more familiar to experienced Final Cut users. They will know about keyframing and color correction and mattes. Here we will be in new territory for iMovie users, and we'll see the application's great power and flexibility come out, multiple video and audio tracks, animation of filters, and customization of effects and generators. Once we're done, whether you're an iMovie or a seasoned Final Cut user, you will know how to work with this amazing application and use its capabilities to the fullest to achieve the results you want, easily and efficiently.

WHAT YOU REALLY NEED

Final Cut Pro requires the latest operating system and computers. It will work only on Intel-based machines and only on computers running Mac OS X 10.6.8 Snow Leopard or 10.7 Lion or higher. The computer needs to have a processor speed of 2GHz or higher and has to have an NVIDIA or AMD graphics card or better. It will run on a computer with the Intel GMA processor shared with main memory; however, this is not recommended as it will limit the application's capabilities. For detailed information on the system requirements for Final Cut Pro, you should check the Apple website at www.apple.com/finalcutpro/specs.

For most people, their needs and their finances almost invariably dictate which computer they purchase for editing video. Generally it's a good idea to get the biggest, fastest, and most powerful computer you can afford. If you have budget constraints, start with an iMac or even a Mac mini. If you need to be on the road a lot, get a MacBook Pro. A MacBook is not recommended for FCPX or its companion app, Motion 5, because the laptop shares memory between user RAM and video RAM. If you have a larger budget, go for it: a 12-core Mac Pro tower, with two 2.66GHz 6-core Intel Xeon processors loaded with lots of RAM.

Multiple Drives

Storage is an essential part of any video system. Digital media file sizes vary enormously. Acquisition formats that are heavily compressed, such as AVCHD, can be as low as about 2MB per second of storage space. Media from a DSLR camera is roughly 6MB per second, or about 350MB per minute, or about 20GB per hour. Higher-resolution formats using codecs such as Apple's ProRes can be about 40MB per second, working out to about 140GB for an hour. As you can see, it won't take long to fill up everything but the largest hard drives. Fortunately, cheap hard drives are available in ever-increasing sizes, with platter speeds, seek times, and caches adequate for working with ProRes material. For higher-quality video using ProRes HQ or large format media, you'll need to have a RAID (Redundant Array of Independent Disks) system. This allows multiple drives to be tied together to work as one unit providing higher capacity and greater speed.

Unless you have a large Mac Pro, you will need external hard drives for your media. Though USB drives are very common, they are not recommended at all for use with video applications. USB drives can be used for archiving your media, but for high-speed delivery of high-resolution high-data-rate media, you

> **NOTE**
>
> Video Codec
>
> Codec stands for "compression-decompression." FCP works using a multimedia architecture that handles video, audio, still images, and other data. This version of FCP supports many different formats natively, though it works best with optimized ProRes QuickTime .mov files. Cameras very commonly shoot in the H.264 codec, which, though it can be used in FCP, is not a particularly good codec for editing. AVCHD, for instance, is MPEG-4 using H.264. This codec compresses video by looking at multiple frames and using the compression information from earlier frames to make subsequent frames even smaller. These formats use GOP (Group Of Pictures) structures to compress the video, using *interframe* compression. These structures are difficult to edit because only some of the frames have complete picture information; your computer processer is needed to re-create "real" frames for editing. I-frame (*intraframe*) compression codecs, such as the ProRes codecs, DV, and the Apple Intermediate Codec, compress each frame as a separate piece of information. This is much easier to edit and requires much less processor power from the computer, allowing the application greater capabilities by not using processor power to decode complex codecs. Some formats such as MPEG-2 used on DVDs are not supported in Final Cut Pro. The MPEG-2 format used on HDV tape can be imported into FCP. DVD media should be converted to QuickTime using MPEG Streamclip (www.squared5.com/). Some card-based AVCHD formats, such as nonrecognizable MTS files from some consumer HD cameras, may need to be "rewrapped" for use on Macs, with utilities like ClipWrap, (www.divergentmedia.com), which also offers the ability to transcode files to various flavors of ProRes422 for editing.

really need at least a FireWire 400 drive and better still a FireWire 800 drive, an eSATA drive, or one of the new Thunderbolt drives that are becoming available. Thunderbolt delivers astonishing speed and performance and works wonderfully for video applications.

It is important that you be aware that only drives that are formatted in Mac OS Extended will work with FCP. They will not work with DOS-formatted or FAT32 or NFTS drives. They simply will not appear in the application.

> **NOTE**
>
> Audio Sample Rates
>
> While video is compressed using codecs, digital audio is recorded by cutting up the sound into samples. The more samples per second into which the audio is divided, the more accurately it will represent the original sound. The standard for almost all digital media is 48kHz (kiloHertz), which means the audio is sampled, sliced into 48,000 segments per second. Older DV cameras still use two sample rate options: 32kHz and 48kHz. The camera manufacturers designate these sample rates as 12-bit for 32kHz and 16-bit for 48kHz. You should always set your camera to record at 16-bit or 48kHz. Higher sample rates such as 96k are also used and are accepted in FCP. Compressed audio, such as MP3 and AAC, can also be used in FCP, but should be avoided. I recommend that compressed audio files be converted to AIFF (Audio Interchange File Format) at 48k. This is an uncompressed audio format, also called Linear PCM in some applications. We'll look at converting audio files in a later lesson.

Because a digital video editing system needs to move large amounts of data at high speeds, you should use separate drives purely for storing video data. You should have one internal hard drive dedicated to your operating system and applications, such as Final Cut Pro, Photoshop or Photoshop Elements, iTunes, your iLife applications, iDVD, iWeb, Garage Band, and so on, and everything else from Internet-access software to word processing and spreadsheets. All of these should be on one drive—your system drive—and all the applications should be in the system's *Applications* folder. All of the Apple applications are usually installed at the top level of the folder. They should not be moved into other folders; otherwise, Software Update or the App Store may not recognize them.

NOTE

Updates

It's a good idea to periodically check the App Store to see whether there have been any updates to the application. Applications are constantly being refined and updated to fix problems or to accommodate developments in hardware or the operating system. The projects used in these lessons may need to be updated to work with your version of the software. The application will update the projects automatically when the new version is launched.

In addition to your internal system drive, you should also have at least one other hard drive that's both large and fast. This drive should carry only your media. A separate drive is much more efficient at moving large amounts of data at high speed. The media drive needs to get that data off the drive very quickly to play it back, especially if there are multiple streams of video and multiple tracks. In addition to the video, the application will often need to play back multiple tracks of audio from various places on the drive simultaneously. That's quite enough work for any one drive to be doing at any one time! Expecting it to access both the application and the operating system can be the straw that breaks the camel's back. You are much less likely to have video playback issues such as dropped frames if you have the media on a separate, dedicated hard drive. This drive should run at no less than 7,200rpm and have at least an 8MB cache. For MacBook Pro or iMac users, external FireWire 800 drives are a good solution, such as those from WiebeTech or G-Tech. With a MacPro, the best solution is to put one or more (up to four additional) large SATA drives inside your computer drive bays, or install an eSATA card to access the speed of eSATA drives, such as those from CalDigit.

Once you have the right computer and hardware, if you haven't done so already, you'll need to download the application from the App Store:

1. Launch the **App Store** from your *Applications* folder. If this is the first time you've used it, you'll have to enter your Apple ID and password, or create one with your credit card information.

2. In the upper-right corner of the App Store is a search box. Type in *Final Cut Pro.*

3. Click on the link to the FCP page or just click directly on the Download button. Apple now offers a free 30-day trial for FCP.

Downloading may take awhile depending on your Internet connection, as it is quite a large application.

If you've downloaded it already, which I suspect you have if you're reading this book, then the first thing you should do after you've launched the application is to go to the **Final Cut Pro** menu and select **Download Additional Content**. This will bring up a dialog that tells you the additional professional codecs that are being installed (see Figures 1.1 and 1.2). iMovie and FCE users will be familiar with the Apple Intermediate Codec and HDV, but the others are professional codecs, including the whole family of ProRes codecs—ProRes 4444, ProRes HQ, ProRes, ProRes LT, and ProRes Proxy. In addition a collection of sound effects and audio plugins is installed.

Once you've downloaded the additional content, you should put the application aside while we look at a few steps you can take to help your computer work with FCP.

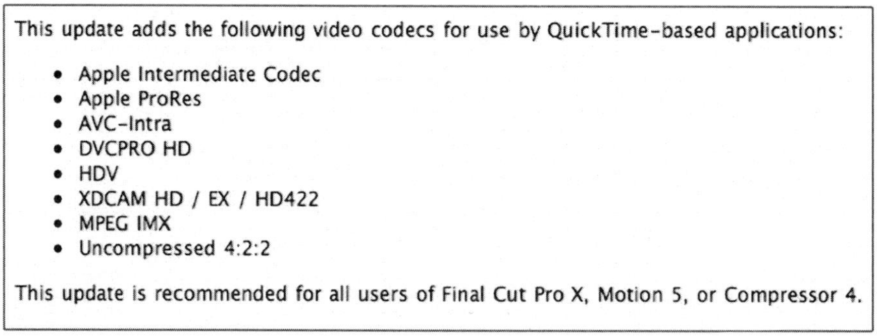

This update adds the following video codecs for use by QuickTime-based applications:

- Apple Intermediate Codec
- Apple ProRes
- AVC–Intra
- DVCPRO HD
- HDV
- XDCAM HD / EX / HD422
- MPEG IMX
- Uncompressed 4:2:2

This update is recommended for all users of Final Cut Pro X, Motion 5, or Compressor 4.

FIGURE 1.1

Additional Video Content.

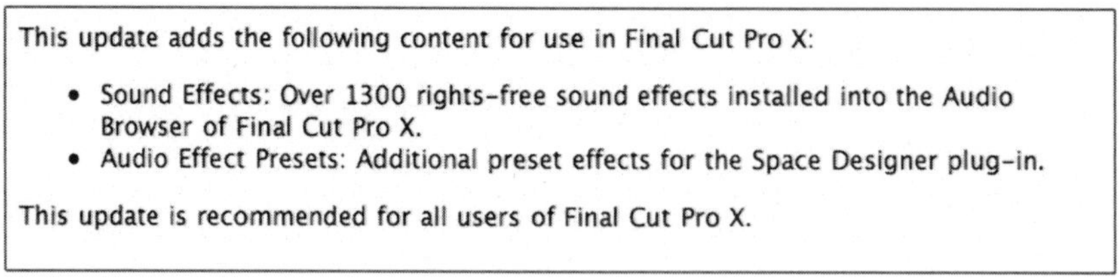

This update adds the following content for use in Final Cut Pro X:

- Sound Effects: Over 1300 rights–free sound effects installed into the Audio Browser of Final Cut Pro X.
- Audio Effect Presets: Additional preset effects for the Space Designer plug–in.

This update is recommended for all users of Final Cut Pro X.

FIGURE 1.2

Additional Audio Content.

> **NOTE**
>
> Audio Contents
>
> Some users have had problems downloading the additional content that comes with FCP, particularly the audio sound effects and loops. If you are having a problem accessing them using Software Update, you can also download them from the Apple website at http://support.apple.com/kb/DL1394.

UPGRADING YOUR OPERATING SYSTEM

Most professionals recommend that when users install a new operating system, such as Mac OS X Lion, they do not simply upgrade it, but rather back up the entire system drive, erase it, install the new operating system, and then download and install the applications. Those of you on Mac Pro towers with older FCP installs under earlier OS X versions and ongoing mission-critical work might consider installing a fresh hard drive in an available bay for just this purpose and then installing FCP on that.

A clean operating system is key to a computer that runs smoothly and efficiently. Of course, with Mac OS X Lion, this is even more problematic than in the past, as the OS is now available only as a download from the App Store. So what to do? It's difficult, but it can and should be done. You will need an 8GB USB flash drive to store your Lion install disk.

1. Begin by downloading the Mac OS X Lion Install application, BUT do not run the installer. If you run the installer, it will upgrade your computer and then disappear from your *Applications* folder.
2. **Control-click** (or right-click) on the downloaded Install application and select **Show Package Contents**. Then go to Contents/Shared Support and find the InstallESD.dmg.
3. Copy this to your desktop and double-click **InstallESD.dmg** to mount it.
4. Plug in your USB flash drive and launch **Disk Utility**.
5. If the flash drive is not formatted for Mac OS Extended, reformat the drive. See the note "Reformatting a New Drive."
6. In the Disk Utility interface, select the formatted USB disk and click the **Restore** tab.
7. You have two boxes here. Drag the mounted Mac OS X Install ESD into the Source box and drag the mounted flash drive disk into the Destination box.
8. Click **Restore**. Wait until it's finished, and you have a Mac OS flash drive installer.

Once that's completed—unless you intend to install on a fresh alternate hard drive—you need to back up your existing system hard drive, usually called the Macintosh HD, though you may well have renamed it. Carbon Copy Cloner is an excellent tool for this. It is shareware and can be downloaded from www.bombich.com. Another useful cloning utility is SuperDuper! from Shirt Pocket, which can be found at www.shirt-pocket.com/. With your data and files and pictures and music all backed up, it's time to erase your drive and install your new operating system.

1. Plug in your new flash drive installer and restart your computer holding the **Option** key.
2. At the beginning of the launch process, you will get the option to launch from the existing system or from the USB flash drive. Select the flash drive.
3. After the system finishes opening, you will get the option to use Disk Utility to erase your hard drive and install the new operating system. That's what you want to do.

Once you have your brand-new, spanking-clean operating system installed, it's time to go to the App Store and get your copy of Final Cut Pro installed. You'll want to install your other applications, especially your professional applications, from their install discs. Then you can use Migration Assistant to help you move your documents and other media, music, and picture libraries back onto your system drive from your backup clone.

NOTE

Reformatting a New Drive

Many hard drives come formatted for Windows. Before you use them, you should reformat them to Mac OS Extended. You do this with Disk Utility.

1. Open **Disk Utility**, which should be in the *Utilities* folder of your *Applications* folder.
2. Select the drive you want to format and click on the **Erase** tab (see Figure 1.3).
3. Set the Volume Format pop-up to **Mac OS Extended (Journaled)**.
4. Name the drive.
5. Click on **Security** and check **Zero Out Data**, as in Figure 1.4. (This figure and others are based on Mac OS X Lion. Your Disk Utility might differ.)
   You only have to zero out on a new drive that has not been properly formatted. If you need to reformat it or erase the drive later, you can do a simple erase that deletes the directory file without having to zero out.
6. For media drives it's recommended that the drive not be journaled. To switch journaling off, select the partition in the Disk Utility, hold the **Option** key, and from the **File** menu, select **Disable Journaling**.

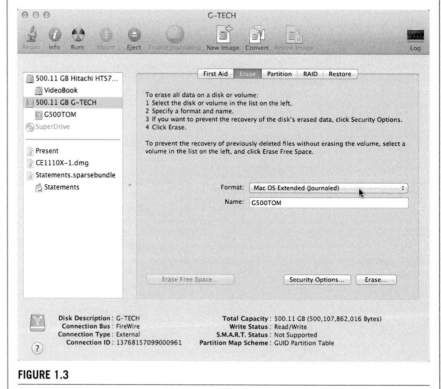

FIGURE 1.3

Disk Utility.

(Continued)

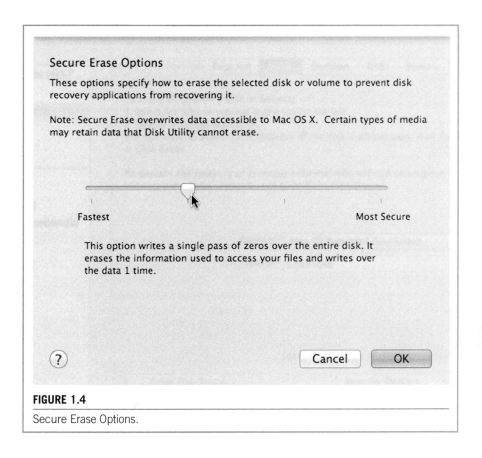

FIGURE 1.4

Secure Erase Options.

TIP

Extra Drive Bay

Out of drive bays on your tower? There's a fifth bay available for a fresh hard drive right under your optical drive bay. The space is normally used for a second optical drive, but with the addition of a pair of 5.25" to 3.5" drive adapter brackets, usually costing less than $5, you have a fifth drive bay in which to install anything you like, and this creates a safe, alternate boot drive on which, after you install Lion on it, you can access with Startup Manager. Invoke your boot-up drive choices by starting your machine while holding the **Option** key for the Startup Manager. This will show you which boot drives are available.

Optimizing Your Computer for FCP

In addition to the hardware, you can optimize your computer for video editing with Final Cut Pro, using a few simple procedures. In **System Preferences**, you can switch off a few items that might interfere with the application's operations while it's running.

For the **Desktop**, most professionals recommend switching off screen savers and working in a neutral, usually medium-gray desktop, because it is more restful for the eyes and does not affect the way your eyes perceive the color of your video. (You'll see this tonal display in the application—mostly gray shades.)

The **Displays** should be set at the resolution settings the system recommends for your monitor, which should be a minimum of 1280 × 768. The application will not run properly at lower resolutions. In Displays the Color tab should be calibrated for use with FCP.

The **Energy Saver** should be set so that the system never goes to sleep. It's less critical, however, that the monitor doesn't go to sleep. I usually set it around 10 minutes, but the system and the hard drive should *never* shut down.

NOTE

Disksomnia

The Energy Saver controls, which you set to not allow the internal drive to shut down, do not control secondary and external drives like FireWire drives. This can be a problem for your external drives; you do not want them shutting down while you're working. If you find your secondary or media drive spinning down or going to sleep (for example, if you take a break), get the simple freeware utility Disksomnia from Martin Baker's Digital Heaven (www.digitalheaven.co.uk/disksomnia) to keep the drive awake while you work. It sends a polling ping to the drive periodically to keep it from dozing off and going to sleep. Digital Heaven is a great resource that provides a number of useful applications and plugins to assist production, including a collection of really useful and pretty transitions, and a storage calculator widget for Dashboard called VideoSpace.

TIP

Automatic Graphics Switching

If you're using a recent MacBook Pro, mid-2010 or later, you'll see an option at the top of the Energy Saver panel that is on by default. This switches between graphics systems used by the computer. It tends to go to the low-powered, lower-performance battery setting to lengthen battery life. If you can, it's better to switch this off while using FCP to improve graphics processing performance.

If you're using a mouse, in the **Mouse** preferences, make sure right-clicking is set up for secondary selection. There are many shortcut menus throughout the application that are contextual. They can be evoked by right-clicking or by **Control**-clicking. In this book I will generally use the term **Control**-click to activate the shortcut menus. Wherever I say that, you can use the right-click to bring up the shortcut menu if you prefer.

I also recommend that you switch off networking. This is done easily by going to **Network Preferences** and creating a new location called **None**. Set up your **None** location without any active connections—no internal modem, no Airport, no Ethernet—everything unavailable and shut off. To reconnect to the network, simply change back to a location from the **Apple** menu that allows access to whatever connection you want to use.

Monitors

Final Cut Pro is designed to work with specific, standard video resolutions. It can edit anything from 640 × 480 standard-definition footage to projects that are 2K or 4K, 2048 × 1024 pixels or 4967 × 2048 pixels, which are feature-film sizes with very wide aspect ratios. For most users, whether you shoot with an AVCHD camera or a DSLR or high-end HD camera, the aspect ratios are constant: All high definition is widescreen; if you work in DV or uncompressed standard definition, your project can be either 4:3 or 16:9.

Apple has done a great deal to make this easier in Final Cut Pro by using the system's ColorSync technology to keep color for your media consistent as it moves between applications. This will keep your video correct on your computer screen, but it will not show you how it would appear on a television set or a video monitor. This is important because your material is being made into a video format, and many video formats are still interlaced and work in color and luminance specifications that date back many years.

While FCP does not support the use of separate broadcast video monitors at this time, it does allow the use of a second monitor to help take some of the pressure off a crowded screen. You may want to use a second computer display to break up the large FCP window. This is really useful for video editing applications. This is especially helpful for long-form video production, or any production with a lot of media, and it's just nice to have the extra screen space. As of this writing, you're allowed only two window items on your second monitor: either the Events Library and Browser or your Viewer.

There are a number of ways to monitor video on your computer, either in the FCP Viewer or in full screen, which iMovie users are probably familiar with, or what was called Digital Cinema Desktop view in Final Cut Express. Full-screen view can be toggled on and off with a button or with a keyboard shortcut. We'll see how to set this up in a later lesson.

In addition, there are other useful tools, such as the Matrox MXO box (www .matrox.com/video/en/products/mac/mxo/fcpx/), which let you put your FCP Viewer on an SDI or an analog monitor, using the computer's FireWire port. If you're working a Mac Pro, there are cards such as the DeckLink series from Black Magic Design (www.blackmagic-design.com) and the Kona cards from AJA (www.aja.com). These can be used to mirror your desktop or as a second display. Some cards also offer a

NOTE

Interlacing

Interlacing is a video system that allows each frame of video to be divided into separate fields; this makes compression of the frame and transmission for broadcast easier, as only half a frame's worth of information is being handled at any one moment. To do this, the frame is divided into two fields made up of thin lines of video that get drawn across the screen. In North America, Japan, and Korea, each field of lines, either all the odd lines or all the even lines, also called upper and lower, are drawn in one-sixtieth of a second. The second field of lines is drawn in the second sixtieth of a second. This produces an image at a frame rate of 29.97 frames per second, often called 30fps, though the true frame rate in North American systems is always 29.97fps. In Europe and much of the rest of the world, the fields occur in one-fiftieth of a second to produce a frame rate of 25fps. For video of a static image, the two fields will be indistinguishable, but if there is motion in the image, the two fields will show a displacement of lines (see Figure 1.5). These are interlacing lines, which are very disturbing when seen on a computer screen, but basic to video displays. You will see these interlaced lines in the QuickTime player and when exporting, but in FCP the default setting does not show interlacing, which can be a problem. If you work with interlaced video, you should switch on the interlaced display in FCP. Many cameras, such as most Sony consumer cameras, still shoot in interlaced formats. Interlaced formats are designated with an *i* as in 1080i60, whereas progressive formats as designated with a *p* as in 720p30. Progressive formats are not interlaced, and each frame is complete and not made up of fields.

FIGURE 1.5

Interlacing.

standard HDMI connector, which carries video and embedded audio to view work on large consumer flatscreens; note, however, that color values will not be accurate broadcast gamut, which requires specialized calibrated TVs.

NOTE

New Drivers

Most third-party hardware technologies will need new drivers to work with FCP. Check on the manufacturers' websites to make sure you have the latest software add-ons that are compatible with FCP, and before you purchase any, make sure that the hardware has been tested and approved for use with the version of Final Cut Pro you're using.

FIRING UP THE APPLICATION

Now it's time to launch that program. Double-click on the icon in the *Applications* folder or, better yet, make an alias in the Dock and click on that (see Figure 1.6).

Launching a new application for the first time is always an adventure, especially when it's as complex as Final Cut Pro, especially one that will now be so unfamiliar to legacy FCP and FCE users. Some software can be intimidating, and some can be downright confusing. When FCP launches, it fills your screen with a single, giant window with lots of panels and buttons to explore. Figure 1.7 shows you the empty FCP window. Notice, because I'm using Mac OS X Lion, there is no visible menu bar. In the upper-right corner is a toggle that allows full-screen mode, a feature that I really appreciate when working in complex applications. The whole interface is a

FIGURE 1.6

The FCP Icon.

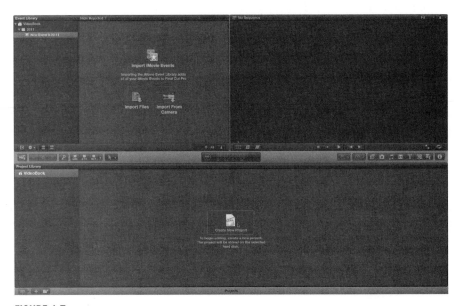

FIGURE 1.7

The Empty FCP Window.

single giant window that fills your computer screen, unlike the FCE interface that had separate windows that could be broken apart. In FCP, like many single-window applications such as iPhoto, closing the FCP window with the red button in the upper-left corner closes the whole application. Again, this is unlike Express, where closing the Browser window also closed the Canvas and Timeline, but left the Viewer and the application itself open.

The Primary Panels

The screen is divided into three primary panels, with two at the top, including a single large Viewer in the upper right, and one across the bottom. The single Viewer is a major change from FCE and previous versions of FCP, where there were separate windows for the Viewer and the Canvas to see the source media and the output.

The **Event Library** is the first panel at the top left of the screen, which is empty except for one event. The Event Library is very similar to the Event Library in iMovie. It is a media browser that gives you access to your drives and the media they hold. The Library must also have one event. If it can't find one, it will make one. Until you make your own, you cannot delete this one. However, you can create a new event, with a new name and specifications and location, and then you can delete the automatically created event, which was generated in a folder in your *Movies* folder inside a folder called *Final Cut Events*.

The Event Library is divided into two panels: the library itself and the **Event Browser**, the empty black area in the middle of the screen, offering you options to Import Files or Import from camera. This is where you can look at the clips you have in your library.

The **Viewer**, the empty monitor on the right, displays the media you're looking at, either being skimmed or played in the Event Browser, or skimmed or played in the Timeline.

The **Timeline** normally occupies the blank area at the bottom of the screen. The Timeline is the place where you lay out your video and audio material in the order you want them. This is where you assemble your project.

The bottom part of the window is now occupied by your **Project Library**. This library is the place where your projects are stored. This is all familiar to iMovie users and a concept that FCE users will need to learn.

Dividing the two halves of the window is the **Toolbar,** which holds a number of buttons and menus as well as the timecode display in the Dashboard.

There are a great many other tools and panels and displays in FCP that are hidden, which we will come to and open as we work with them.

THE PROJECT FACADE

To some extent an FCP project is simply a facade. In other words, whatever is in the project and whatever you bring into the project—the media you import—are equivalent to the aliases for your media. While you are working with these aliases, you are using them to pass instructions to the computer about which pieces of video and audio to play when and what to do with them. The conveniences created for you in the application are an elaborate way of telling the computer what to do with the media on your hard drives and how to play it back. All of the clips in the project, whether in the Event Browser or the Timeline, are simply pointers to the media on the hard drive. This is a nondestructive, completely nonlinear, random-access artifice. This means that your media is not modified by anything you do in the application, but you can arrange the clips and work on any portion of your project at any time. It also means you can access any piece of media from anywhere on your hard drive at any time. The clips are not brought into the project or placed in the Timeline. They never leave their place on the hard drives. You can change the names in the browser to anything more convenient, and this change has no effect at all on the data stored on your hard drive. On the other hand, if you change the names of the clips on your hard drive, that *will* confuse FCP.

Window Arrangements

The panels can be altered from the default state to preset configurations. As in iMovie, the little button in the extreme lower left of the Events Library (see Figure 1.8) will close the Event Library, giving expanded space to the Event Browser. This is a toggle button that will open and close it as needed. You can also open and close the Event Library with the keyboard shortcut **Shift-Command-1**.

FIGURE 1.8

Closing the Event Library.

This is the first of many, many keyboard shortcuts scattered throughout this book. I cannot emphasize enough how important it is to learn and use keyboard shortcuts when working with complex applications such as FCP.

> **NOTE**
>
> Keyboard Shortcuts
>
> Using any complex application like Final Cut Pro or Photoshop or Motion can be done much, much more efficiently from the keyboard than with a mouse. FCP is significantly more mouse driven than earlier versions of the application in which you could almost entirely drive the application from the keyboard. You will be using the mouse and trackpad more with FCP, so use the keyboard whenever you can. A few days of working solely with a mouse will send you on your way to Carpal Tunnel Syndrome and a whole world of pain. Your wrist will thank you for it every day you can spend more time on the keyboard and less on the mouse. The FCP keyboard is pretty comprehensively mapped, and it would be very beneficial to learn the most important shortcuts. A complete list of shortcuts can be found in the application's Help menu. If trying to remember all of the keyboard shortcuts is shorting out your brain, you can get color-coded special keyboards with keys that display the topmost shortcuts. A great tool is Loren Miller's KeyGuide, a laminated color-coded keyboard display of all FCP's shortcuts. No FCP editor should be without one. Miller makes them for a number of applications, including the new Final Cut Pro. You can find out more about them and order them from www.neotrondesign.com.

In addition to the Event Library, another important panel on the right is the Inspector, which can be opened and closed with a button. That's the little *i* button in the extreme right of the Toolbar (see Figure 1.9) that opens and closes the Inspector, which can also be toggled with the keyboard shortcut **Command-4**.

At the extreme lower left of the window is the button that closes and reveals the Project Library (see Figure 1.10). When the Project Library is closed, you have access to the Timeline panel where your editing is done.

In FCP, you can resize panels in the window dynamically by grabbing the edges where the mouse pointer changes to a Resizing tool (see Figure 1.11). When you pull with the **Resizing** tool, the panels will move, expanding and contracting as needed to fill the available space. The Timeline/Project Library can be dragged vertically. The Viewer, the Event Browser, and Event Library can be resized horizontally.

FIGURE 1.9

Opening the Inspector.

FIGURE 1.10

Closing the Project Library.

FIGURE 1.11

The Panel Resizing Tool.

You can always revert to the original window layout from the menus by using **Window>Revert to Original Layout**. Unlike Final Cut Express, FCP has no keyboard shortcut for this, but if you want, you can customize the keyboard to set it to the same keyboard shortcut **Control-U**. We'll look at customizing the keyboard in a later lesson.

SUMMARY

So ends Lesson 1. This chapter is designed to help you to get started correctly in using Final Cut Pro, making sure your system is set up to edit digital video with a professional application. The hardware requirements for the computer and the graphics card are pretty stringent and need to be adhered to closely. Once you have the operating system and the application installed and up to date, you need to ensure that your drives are capable of handling the requirements of high-data-rate media. Once everything was set up, we launched the application and looked at the full-screen, single-window interface. We looked briefly at the interface and its separate panels. In the next lesson, we'll look at the interface in greater detail with an existing project.

The Final Cut Pro Interface

2

LESSON OUTLINE

Now that you've set up your system, let's look more closely at each of the components of Final Cut Pro's interface and each of its panels. To do so, you must open a project that has something in it, and to do that, you need to download the project files and media from the book's website if you have not done so already. Go to www.fcpxbook.com. On the website there are a number of ZIP files, each of which contains a sparse disk image that we will work with. You will probably need to download all of them over the course of the book, but for now you simply need the first ZIP file called *FCP1.sparseimage.zip*. This is a fairly large file, about 500MB in size, so downloading will take some time. The sparse image itself when it's unzipped is even larger. The other ZIP files contain other sparse images that contain other projects and media files.

A sparse image is a type of disk image that has a maximum allowable space. In the case of *FCP1*, the maximum size is 10GB. The image itself is nowhere near that large, but it has the capacity to expand to that size. Because the sparse images are expandable, you need to have plenty of free drive space on any drive that holds these images. When the *FCP1.sparseimage* is mounted, it appears on your system as just another hard drive. FCP treats it like any other hard drive and gloms onto it. As soon as the application detects a drive, it searches it for the folders it uses. It looks specifically for two folders, one called *Final Cut Events* and the other called *Final Cut Projects*. The application looks only at the top level of any mounted drive or disk image, or in the user's home *Movies* folder. If those folders are not there, the application will not recognize the drive.

> **NOTE**
>
> Most of these sparse images were generated with events and projects from FCPX 10.0.1. They won't open in version 10.0; a dialog will appear indicating "the catalog is too new." The update is free if you purchased FCP at the App Store.

LOADING THE LESSON

Once you have downloaded the ZIP file from the website, you should double-click it to unzip the sparse image. Next, you should move the *FCP1.sparseimage* to your media drive before opening it. When you copy or move it to the media drive, the contents will play from that drive. FCE users are accustomed to keeping the project files on the system drives, but this is not the case with FCP. Project files are kept inside the *Final Cut Projects* folder. This folder contains the render files associated with the projects. These are large, high-resolution files using the ProRes codec. Because of this, both the *Projects* folder that holds your projects and the *Events* folder that holds your media need to be on your dedicated media drive.

1. Once you have the sparse image where you want it, double-click it to mount the disk.
2. Launch Final Cut Pro, which will detect the events and project folders in the mounted disk and display them in the Event Library and Project Library.

The media files that accompany this book are heavily compressed H.264 files. Playing back this media requires a fast computer. If you have difficulty playing back the media, you should transcode it to proxy media.

1. To convert all the media, select the event in the Event Library, **Control**-click on it, and choose **Transcode Media**.
2. In the dialog that appears, check **Create proxy media**.
3. Next, you have to switch to proxy playback. Go to **Preferences** in the **Final Cut Pro** menu.
4. On the **Playback** tab, click on **Use proxy media**. Be warned: If you have not created the proxy media, all the files will appear as offline. In this case, you will have to switch back to **Use original or optimized media**.

THE EVENT LIBRARY

If you have the Event Library open, in the upper left you will see the *FCP1* disk and inside it the Event it has in its *Events* folder. If the disk is closed, click the disclosure triangle next to it to open it, and it will show the Event it holds. This is usually displayed in date order. If it's hidden, click the disclosure triangle next to the year to open it. If the event is closed, click the disclosure triangle next to it to see the collections available in the event. The clips display in the Event Browser to the right of library (see Figure 2.1, also in the color section).

TIP

Media Offline

Should you open the disc image and the content displays with icons that are red with a yellow warning exclamation mark, go to your FCP **Preferences** and make sure that the Playback tab has **Playback** set to **Use original or optimized media** and is not set to **Use proxy media**. We'll look at Preferences in more detail in the next lesson.

FIGURE 2.1

Event Library and Browser.

WHAT'S AN EVENT? WHAT'S A PROJECT?

Events and projects are familiar to iMovie users but probably very new to Final Cut Express and Pro users. The name *event* might be confusing for some, but try not to focus on the word. Think of it as your production, whether it's a single event like a wedding, or an extended event like a production that goes on over many weeks or months.

The event holds your media. It's the equivalent of the capture scratch folder in Express and legacy FCP. This is the folder in which your media is located on your hard drive, and as such it needs to be on a big, fast, dedicated media drive. In the Event Library, each event can be viewed as a container in which you organize clips. The event in the Event Library is similar to but not analogous to the old Project Browser. In simplified form, the old Project Browser becomes the Event Browser. This is a very loose comparison, as these applications function very differently. It is probably better not to try to think in terms of equivalences, but to treat this as a wholly different way of working.

If events are confusing for Express and legacy Final Cut Pro users, then projects might be even more so. The old applications used a Project Browser as a list of video and audio clips, graphic files, saved titles and effects, and multiple sequences in which your program was edited. In FCP X, projects are sequences, contained in the Project Library. There is no overarching single project that contains your sequences. Instead of one center of attention, you have two. The master container for all your media is the Event Library. The master container for all your editing work is the Project Library; double-click any project and you're taken to the Timeline.

TIP

Launching the Application

Many legacy users would boot up their application by double-clicking their current project icon instead of the application icon. You should not do this in FCP. Instead, simply launch the application. Each project folder has a *CurrentVersion.fcpproject,* and each event folder has a *CurrentVersion.fcpevent* icon. These should not be opened, and the names should never be changed.

TIP

Control and Manage Your Events

FCP wants to display all projects and events all the time, but only when they're in specific locations. Often you don't want to see events or projects, or you don't want others to see what other projects you're working on. You could manually move projects and events around at the Finder level to hide them from FCP, but there's a very handy, inexpensive tool to do this. It's called Event Manager X and can be found at http://assistedediting. intelligentassistance.com/EventManagerX/.

Each project in the new application is a separate sequence. If you need to make multiple projects for a production, for backup, to try different edits, or for different versions, you work with multiple projects and group the projects together in a folder.

FIGURE 2.2

Project Library.

The folder structure is echoed at the Finder level as well. All your projects are inside this folder and can be named and organized as you wish, just as in the legacy software.

Any media in any event can be used in any project, though in practice it is probably a good idea to associate one or a number of grouped events to a single project. That's the way this event and project are laid out. The event *JungleGarden* holds the media for the project *JungleGarden*.

The Event Library is the place where every production edit begins. This gives you direct access to the media on your hard drive, whether it's your media drive or a mounted disk image. The Event Library and the Project Library below it (see Figure 2.2) are direct windows to your hard drives; they hold all your media and your projects, respectively. Inside the Project Library, your project *JungleGarden* is your edited program. The Event Library event *JungleGarden* holds the media and is made up of organizational items, folders, and collections—either Keyword Collections or Smart Collections. Your media, as seen here, is made up of video clips but can also be audio files or still images or graphics files created in Photoshop or other applications.

Smart Collections are an important new feature in FCP. They are designated by a purple file icon with a gear on it. Like smart albums in iPhoto and Aperture and smart playlists in iTunes, Smart Collections are designed to automatically update as new media is added to the event that meets the Smart Collections criteria. We'll look at collections in greater detail in Lesson 4 on organizing.

Keywords are not new to experienced iMovie users. Keyword Collections are created by applying keywords to clips or even to portions of a clip. These collections are designated with a pale blue icon with a key on it.

In this event I have put the two collections inside a folder, though this is not necessary, but it is another of the organizational tools available to you that we will look at later.

Event Library Buttons

There is a cluster of buttons in the lower left of the Events Library that control it (see Figure 2.3). The first on the left simply opens and closes the panel accessing the Events Library, making more space available for the Event Browser. You can also do this with the keyboard shortcut **Shift-Command-1**. The gear icon to the right of it

FIGURE 2.3

Event Library Buttons.

FIGURE 2.4

Event Library and Browser Settings.

accesses the settings for the library and for the Event Browser (see Figure 2.4). Here you can set the appearance of the library and how items are grouped and displayed. The last two items in that menu control the display in the Event Browser. These selections can also be made in the menus using **View>Event Browser**.

There are simple keyboard shortcuts to select each of FCP's panels. The principal panel shortcuts are shown in Table 2.1.

Table 2.1 Principal FCP Panel Shortcuts	
Panel	**Shortcut**
Events Browser	Command-1
Toggling the Event Library	Shift-Command-1
Timeline	Command-2
Timeline Index	Shift-Cmmand-2
Viewer	Command-3
Inspector	Command-4
Effects Browser	Command-5
Color Adjustments	Command-6
Video Scopes	Command-7
Audio Adjustments	Command-8
Background Processes	Command-9
Project Library	Command-0
Cycling clockwise through the panels	Tab
Cycling counterclockwise through the panels	Shift-Tab

THE EVENT BROWSER

The **Event Browser** is the panel in the middle of the FCP window (see Figure 2.5). It is much like the Event Browser in iMovie and holds much of the same information as the Final Cut Express and earlier Final Cut Pro Browser did. It also combines some of the functions of the Viewer from those earlier applications, making the Event Browser a key part of the application. Much of the selection and editing of clips in Final Cut takes place in the Event Browser. This is the place where you manipulate individual clips, mark where you want them to start and end, and prepare them for your project. To access the clips for your project, select one of the collections in the Event Library, or to see all of the clips in the event, select the event itself.

Event Browser Controls

There are two buttons for the Event Browser in the lower left, next to the gear pop-up menu. These buttons let you switch between Filmstrip view, the left of the two buttons, and List view, the right of the two buttons. Switching between these views can also be done using **View>Event Browser** and setting either List or Filmstrip. Or you can use the keyboard shortcuts **Option-Command-1** for Filmstrip view, which is the default display, and **Option-Command-2**, which is more like the Express Browser list view.

At the top of the browser are two other controls. The pop-up on the left, which defaults to Hide Rejected, lets you select what clips you see in the panel (see Figure 2.6).

FIGURE 2.5

Event Browser.

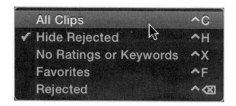

FIGURE 2.6

Event Browser Selection Pop-up.

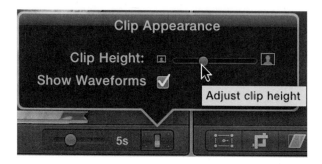

FIGURE 2.7

Filmstrip Height and Waveforms.

Notice the keyboard shortcuts that appear in the menu. The selections range from All Clips, which I use most of the time, to showing just clips that haven't been rated or keyworded, to Favorites, and to displaying Rejected clips and to Hiding Rejects.

On the right is a search box that lets you looks for clips in the browser based on text and other clip criteria. This is also the basis for creating Smart Collections, which we'll see in detail in Lesson 4.

Filmstrip View

When you switch to Filmstrip view (**Option-Command-1**), two items are available in the lower right of the panel, a slider and a switch button, which are not available when in List view. There are a few other of these switch buttons around the application, and they control the appearance of the window. If you click on the switch, a pop-up menu appears. Inside is a slider that adjusts the clip height, and a checkbox that allows you to toggle the audio waveforms off and on (see Figure 2.7).

The slider next to the switch lets you change the view of the clips (see Figure 2.8). It controls how many thumbnails are used to represent your clip. When it's set to the right to All, each thumbnail in the browser represents a single clip. As you move the slider to the left, the display changes to show the clips with filmstrips—first one thumbnail for 30 seconds, then 10 seconds, then 5, 2, 1, down to one thumbnail for half a second of video.

> **TIP**
>
> Switch to All
>
> If the Event Browser is the active panel, the keyboard shortcut **Shift-Z**, which in the Viewer and the Timeline is Zoom to fit the panel, will switch the filmstrip display in the browser to All.

List View

Use the View menu or the List view button or press **Option-Command-2** to switch the Event Browser to list view (see Figure 2.9). Here you see the clips as a list, with a filmstrip view of the selected clip at the top of the window. Audio files would show a waveform in the preview area, while a still image or graphic will display the media at the top of the list.

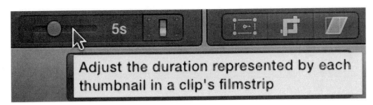

FIGURE 2.8

Filmstrip Thumbnail Control.

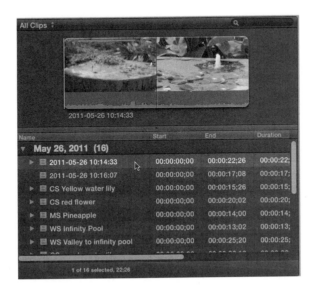

FIGURE 2.9

Event Browser List View.

In the Event Browser you can change the names of your clips, as I have done for most of these clips. All clips from AVCHD, which this media is, come in renamed with a date/time stamp of when it was recorded. To rename a clip, select the clip and press the Return key. This selects the name and lets you enter a new name.

1. Select the first clip by clicking on the name line as in Figure 2.9.
2. Press the **Return** key to select the name.
3. Type in *Small pool push to fountain*.
4. Select the second clip and name it *Lily pond tilt down and up*.
5. Click the **Name** header at the top of the list to reorder the clips alphanumerically based on their name.

Renaming the clips only changes their name in the Event Library and does not affect the name of the media on your hard drive. To see this, either **Control**-click on the clip name in the list (or on the preview at the top if you prefer) and select **Reveal in Finder** from the menu. You can also use the keyboard shortcut **Shift-Command-R**. This will take you to the media file on your hard drive, which is in the *Original Media* folder inside the *JungleGarden* event folder, which is inside the *Final Cut Events* folder.

TIP

Control-Clicking

If you do not have a three-button mouse, either an Apple Magic Mouse or a third-party mouse, you really should get one. If you have the Apple Magic Mouse, make sure you go to the **System Preferences** and enable right-clicking in the Mouse controls. Any inexpensive USB three-button mouse with a scroll wheel will also do. The application uses **Control**-clicking to bring up shortcut menus, and the properly configured scroll wheel can move sliders as well as window displays. If you are using a three-button mouse, whenever I say "**Control**-click," you can press down the right side of the mouse while clicking to bring up the shortcut menu. On a laptop without a mouse, press the **Control** key and mouse down to bring up the shortcut menu, or use the gesture with the trackpad.

Viewing Clips

Let's look at our clips. There are a number of different ways to play your video. If you select a clip, the preview at the top shows you a short filmstrip preview. The number of frames that appear depends on the width of your monitor and the space allowed for the Event Browser, so on a long clip the frames might be far apart, whereas on a shorter clip they'll be closer together.

One way to view the video is very familiar to iMovie users but completely new to legacy Final Cut users, and that's skimming. Simply move the pointer across the video preview at the top of the screen, and the content will display in the Viewer on

FIGURE 2.10

The Skimmer and Playhead.

the right. You don't need to press the mouse to drag across the screen. Simply slide the pointer over the screen. Where you are in the preview is indicated by the thin, pink, vertical bar called the **skimmer** (see Figure 2.10, also in the color section).

There are two separate indicators to show where you are in a clip. One is the skimmer, and the other is the **playhead**, the gray line in Figure 2.10. The skimmer floats wherever the pointer is, while the playhead remains in a specific location until you actually play the clip. When you press the Play button at the bottom of the Viewer or use the keyboard shortcut, the playhead moves together with the skimmer. When you move the pointer around the interface, the skimmer moves. If you move the pointer out of the preview window, the skimmer disappears and the Viewer shows the playhead position. If you want to move the playhead to where the skimmer is located, click on the clip. The skimmer is a wonderful tool that lets you move around your material very, very quickly. This tool is, of course, old hat for iMovie users, but takes quite a bit of time for Express users to get used to. However, I think the time invested in getting comfortable with this tool is well worth the effort.

Skimming is a really fast way to find media in your browser. You can quickly skim through everything in an event or a collection without having to open individual clips. They're all there for you to see and access.

Sometimes you might want to skim video without hearing the audio. You can toggle this on and off from the **View>Audio Skimming** or use the keyboard shortcut **Shift-S**. There is also a button in the Timeline for this, which we will see in a moment.

You're also going to want to play the video in realtime. To play the clip, click in the preview area and press the Play button at the bottom of the Viewer. If you like working with the mouse, this is just for you, but it is not the most efficient way to work by any means. Remember the keyboard is your friend. Learn how to use the keyboard because it makes it much easier to control your editing than using the mouse.

To play the video, press the **Spacebar**. The clips will start playing from wherever the playhead or skimmer is. To pause, press the Spacebar again—**Spacebar** to start, **Spacebar** to stop. To play the clip backward, press **Shift-Spacebar**. This method is much quicker, and it keeps your hands on the keyboard and off the mouse.

Another common way to play the clips is with the **L** key: **L** is play forward, **K** is pause, and **J** is play backward. On your keyboard, the keys are clustered together, but you're probably thinking, "Why not comma, period, and slash?" There is a method to the madness. **J, K,** and **L** were chosen because they're directly below **I** and **O**, which are used to mark the In and Out points, the beginning and end of selections on clips. These are also probably the most frequently used keys on the editing keyboard. Hence, **J, K,** and **L** are positioned conveniently for the finger of your right hand with the **I** and **O** keys directly above them.

You can view your video at various speeds. You can fast forward by repeatedly tapping the **L** key. The more times you press **L**, the faster the clip will play. Similarly, tapping the **J** key a few times will make the clip play backward at high speed. These are your VCR controls.

To play a clip one frame at a time, tap the **Right** arrow key. To play it slowly, hold down the key. To play slowly backward, hold down the **Left** arrow key. To jump forward or backward in 10-frame increments, use **Shift** with the **Left** or **Right** arrow keys. Pressing **K** and **L** together will also give you slow forward, and **K** and **J** together will give you slow backward. To go to the beginning of the clip, press the semicolon key (;), and to go to the end of the clip, press the apostrophe key ('). Use the **Down** arrow key to go the next clip in the list and the **Up** arrow key to go to the previous clip in the list. The **Home** key (**fn-Left** arrow on many keyboards) takes you to the first clip in the list, and the **End** key (**fn-Right** arrow) takes you to the last clip in the list. These arrow keys also work when the Event Browser is in Filmstrip view. A list of important keyboard shortcuts can be found in Table 2.2.

Another useful keyboard shortcut is the forward slash key (/), which will play through a selection from beginning to end.

Event Browser Details

If the Events Library is open, close it with the button in the lower left of the Events Library panel or press **Shift-Command-1**, giving your Event Browser more space. With the Browser in List view, stretch out the Browser window to the right, and you will see just some of the many things the Browser displays in List view. If you **Control**-click in the Browser list header, you will get further options that are available (see Figure 2.11).

Table 2.2 Some Principal Keyboard Shortcuts

Play	L
Pause	K
Play backward	J
Fast forward	Repeat L
Slow forward	L and K or hold Right arrow
Fast backward	Repeat J
Slow backward	J and K or hold Left arrow
Forward one frame	Right arrow
Backward one frame	Left arrow
Forward 10 frames	Shift-Right arrow
Back 10 frames	Shift-Left arrow
Go to previous clip	Up arrow
Go to next clip	Down arrow
Go to first clip	Home
Go to last clip	End
Mark the In point	I
Mark the Out point	O
Go to In point	Shift-I
Go to Out point	Shift-O
Play selection	/
Select Clip	X
Add Marker	M

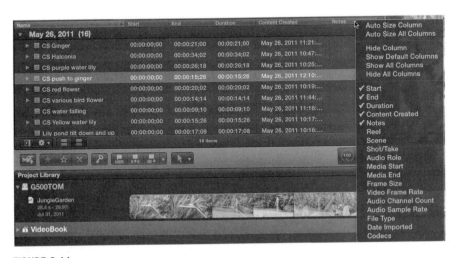

FIGURE 2.11

Browser Lists.

TIP

Ordering

You can arrange the order in which clips are shown in List view by selecting the column header. By clicking the little triangle that appears in the header, you can change the order from descending to ascending.

The Browser shows the duration of clips and the Start and End times. You also see a box of Notes, as well as technical information, though this is best viewed in the Inspector as we will see.

In addition to adding more column information to the Browser, you can move any of the columns by grabbing the header at the top of the column and pulling it to wherever you want the column to appear. So, for instance, you could drag your Notes column next to your Name column. Only the Name column cannot be moved; it stays displayed on the left side of the panel. The Notes column allows you to add descriptions and other information about your clip.

TIP

Tab Key

If you have a clip selected, the **Tab** key will, like the **Return** key, select the clip name to change it. If you press **Tab** again, the cursor will move to the Notes column, where you can enter information about your clip, such as a few words about an interview. If you select sections of a clip using keywords, each keyword or favorite selection will have its own Notes box.

THE VIEWER

The Viewer in FCP is your window on your video and shows whatever clips you're playing back, whether skimming or playing in the Event Browser or in the Timeline. It switches to whichever is the active window. The controls centered under the Viewer (see Figure 2.12) are very simple. There's a Play button to play the video from wherever the playhead is parked in whichever window is active, the Event Browser, the Timeline, or even the Project Library. To the left of the Play button are two buttons; the leftmost steps backward in one-frame increments (**Left**-arrow), and the other steps forward in one-frame increments (**Right**-arrow). To the right of the Play button are two buttons; the one closest to the Play button is used to move to the beginning of the selected clip (the **;** key**),** and the one farther away is used to move to the end of the clip (the **'** key).

On the far lower right of the Viewer are two buttons (see Figure 2.13). The one on the left displays your Viewer in full-screen mode, which can also be evoked with the keyboard shortcut **Shift-Command-F**. While in full-screen mode, all your standard keyboard shortcuts work—**Spacebar**, **JKL**, **I** and **O**, and so on. To get out of full-screen mode, press the **Escape** key.

Next to the full screen is a button that turns on looped playback, which will play any selection, clip, or even an entire project again and again. It can be toggled on

FIGURE 2.12

Viewer Controls.

FIGURE 2.13

Full Screen and Looping Buttons.

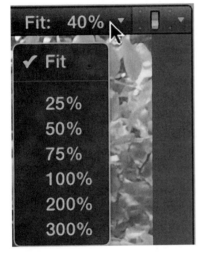

FIGURE 2.14

Zoom Pop-up.

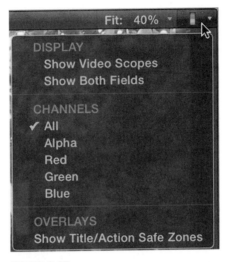

FIGURE 2.15

Viewer Appearance.

and off with the shortcut **Command-L**. The looped playback button turns blue when looping is active.

In the lower left of the Viewer are three buttons for image manipulation: transformation, cropping, and distorting. We'll look at these controls later in the book in Lesson 12.

The name displayed in the upper-left corner is the name of the clip or project that the Viewer is playing back. The icon next to it indicates whether the item playing is in an event, a star icon, or is a project, a document with the FCP icon on it.

In the upper right of the Viewer are two important controls. The one on the left is the Zoom pop-up (see Figure 2.14) that lets you set the display size of the video in the viewer. It's normally set to Fit, which can be evoked with the shortcut **Shift-Z**.

Next to the Zoom pop-up is another toggle switch similar to the one at the bottom of the Event Browser in Filmstrip view that controls the appearance of the window (see Figure 2.15). **Show Both Fields** will display any interlacing in your video. You can also display various video channels, **All**, or just specific red, green, and blue channels rendered in grayscale values, or the alpha (transparency) channel. If there is no transparency in the image, the screen will appear completely white. Finally the pop-up menu can turn on your **Video Scopes** and display title and action-safe areas, which we'll look at when we do titling in Lesson 9.

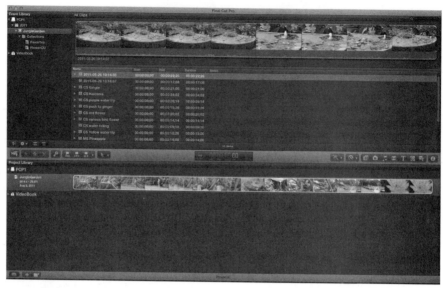

FIGURE 2.16

FCP Without the Viewer.

In addition to seeing the Viewer within the FCP window and viewing its contents in full screen, you can also place the Viewer on a second computer monitor using **Window>Show Viewer on Second Display**. If you do that, your window will look like Figure 2.16, giving you more space for Browser columns and for the Inspector, which we will see in a moment. You can also put your Events Library and Browser on the external monitor as in Figure 2.17, leaving the primary FCP window for the Viewer and the Timeline.

> **NOTE**
>
> Fixed Window
>
> Be aware that you cannot resize the Viewer window; it takes up a full screen on your second monitor. Also be aware that you must return to the Window menu to re-integrate the Viewer to your main display; Escape will not work. You may, of course, map an unused key for a keyboard shortcut to the Show/Hide Viewer on Second Display function, such as **Control-W**, which will toggle it. We'll look at customizing the keyboard in a later lesson.

THE INSPECTOR

One of the most important panels in FCP is hidden from view when the application is first opened, and that's the Inspector. You can call it up from the **Window** menu or using the keyboard shortcut **Command-4** or by clicking the **i** button for the Inspector

FIGURE 2.17

Event Library on a Separate Monitor.

FIGURE 2.18

Inspector Button.

on the far right of the Toolbar (see Figure 2.18). The Inspector is contextual and will change to display information and controls for whatever is selected. It can provide information about a selected clip and give controls for clip transformations, color correction, events, transitions, and titling tools.

If you have a clip selected in the Event Browser, you are not limited to the few items that appear in the default list view. For a Browser clip, the Inspector has three tabs at the top. There is a Video tab, which has limited controls until the clip is placed in the Timeline. The Audio tab (see Figure 2.19) gives you Volume control, Pan control, and Channel Configuration. Changing these prior to editing the clip into the Timeline will set these parameters for every instance of the clip you use. Of course, they can be changed again

FIGURE 2.19

Inspector Audio Tab.

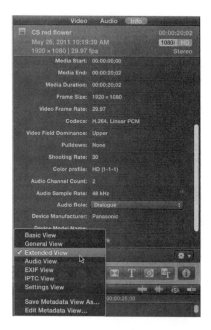

FIGURE 2.20

Inspector Info Tab.

in the Timeline whenever you like. The Info tab (see Figure 2.20) provides extensive information about the media specifications, as well as provides multiple types of information, much more comprehensive than in any other application. In addition, with the gear pop-up menu in the lower-right corner of the Inspector, you can add custom metadata and customize your naming conventions for clips (see Figure 2.21). Information and organization are key components of this application, as we will see.

The further we go into the book, the more important the Inspector will become.

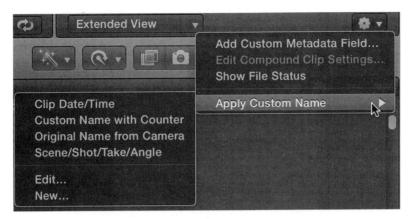

FIGURE 2.21

Inspector Customization.

FIGURE 2.22

The Toolbar.

FIGURE 2.23

The Dashboard.

TOOLBAR

Across the middle of the FCP window, dividing the Event Browser and Viewer from the Project Library or Timeline, is the Toolbar, which is divided into three sections, with buttons on the left and right and the Dashboard in the center (see Figure 2.22).

Dashboard

The Dashboard (see Figure 2.23) shows the timecode for whatever clip is being played back, whether in the Event Browser or the project itself. It switches instantly from one to the other as the emphasis is changed. The Dashboard is also a tool. It can be used to numerically move clips and edits, and by clicking the timecode or pressing **Control-P**, moving the playhead itself. It can also be used to change the duration of a clip by double-clicking the timecode or pressing **Control-D**.

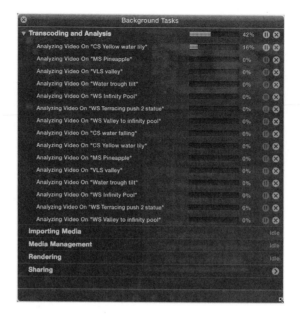

FIGURE 2.24

Background Process HUD.

On the left side of the Dashboard is a circle that will display the time remaining for any background process like rendering or analysis. Clicking the circle or pressing **Command-9** will evoke the Background Process HUD, which provides a detailed display or the functions being undertaken (see Figure 2.24). **Command-W** will close the HUD.

On the right of the timecode display are the tiny audio meters. They're not much use as meters except as a confidence check, but by clicking on them or pressing **Shift-Command-8**, you can bring up a full set of meters that can display either two bars for stereo audio or the six bars for 5.1 Surround if your project is set to that specification (see Figure 2.25).

Left-Side Buttons

The first button on the far left side is the Camera Import button (see Figure 2.26), which allows you to bring in media from your camera or other device like a card reader or the card slot on your computer or from a camera archive. We'll look at this in the next lesson.

Next to that is the cluster of three buttons for favorites and rejects and unrating, together with the button for adding keywords (see Figure 2.27). These are organizational tools that we'll look at in Lesson 4.

To the right of these are three edit buttons and a pop-up (see Figure 2.28), which allow you to do connect, insert, and append edits. We'll use these tools in Lesson 5.

FIGURE 2.26

Camera Import Button.

FIGURE 2.27

Favorites, Unrated, Rejects, and Keywords.

FIGURE 2.28

Edit Buttons.

FIGURE 2.25

Audio Meters.

To the right of that is the Select tool pop-up (see Figure 2.29). The Select tool can be evoked with the **A** key, and the other trimming and control tools that appear in the pop-up beneath it can all be called up with single-letter shortcuts. These tools we'll use in Lesson 6.

Right-Side Buttons

iMovie users will be very familiar with many of the buttons on the right side of the Toolbar. They have similar functions, accessing titles, transitions, still images, and effects. The first two buttons will be new, however. The Enhancements button with the magic wand (see Figure 2.30) is used to access enhancement features for audio and video. Next to that are the retiming controls (see Figure 2.31) that allow you to change the speed of clips in your projects. These will all be used in later lessons.

FIGURE 2.29

Select and Trim Tools.

FIGURE 2.30

Enhancements Menu.

The buttons next are those familiar to iMovie users (see Figure 2.32), and each opens a panel that gives you access to effects for both video and audio. The still camera button lets you add still images from your iPhoto and Aperture libraries to FCP. The musical note takes you to your iTunes library as well as to loops, sound effects, jingles, and other audio content. The hourglass icon opens the transitions browser. The T button gives access to FCP's 159 titling tools, most of which are animated in some way. The countdown clock lets you access the Generators panels, which Final Cut Express users will be familiar with. And finally, the Theme button holds groups of transitions and text tools, which are designed to work together.

FIGURE 2.31

Retiming Menu.

FIGURE 2.32

Effects and Other Panels.

THE PROJECT LIBRARY

The area at the bottom of the screen can be occupied by one of two separate panels: the Project Library or the Timeline. When we first launched the application and looked at the contents of the FCP1 sparse image, we saw the Project Library (see Figure 2.33), with the Timeline panel closed. As with the Event Library, you can have multiple projects on multiple drives and can access any of them from the library. You can create a project on any local drive, internal or external, and you can organize your projects in folders. Every drive has a disclosure triangle next to it if it has a Final Cut events or projects folder on it. The triangle allows you to hide and reveal the FCP contents of that drive.

In the Project Library, the project can be skimmed, viewed (even in full screen), or exported directly from the library without opening the Timeline. Just as in the Event Browser, the common keyboard shortcuts, **Spacebar,** and **JKL** will play, rewind, and fast forward your project in the library.

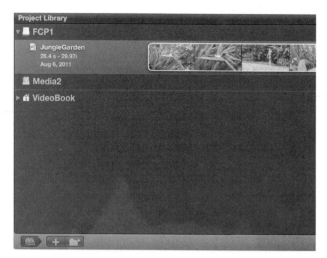

FIGURE 2.33

Project Library.

FIGURE 2.34

Project Library Buttons.

To open an existing project, you first have to select it; then you can either

- Double-click the project filmstrip
- Press the **Return** key
- Press **Command-0** (**zero**), which toggles the Project Library open and closed
- Or, click the film reel icon in the lower left of the panel (see Figure 2.34)

You can have as many projects as you want or as your drive space will allow. You can copy a project from one drive to another simply by dragging it to the drive icon in the Project Library. You can also move a project to another drive by either dragging while holding the **Command** key or by using **File>Move Project**. If you move the project, you will get the dialog in Figure 2.35. This allows you to select the drive you want to move the project to, as well as whether to move just the project or to move the referenced event folder also.

You can also duplicate a project by pressing **Command-D**, by using **File>Duplicate Project**, or by **Control**-clicking the project and selecting **Duplicate Project** from the shortcut menu. If you duplicate a project, you get a

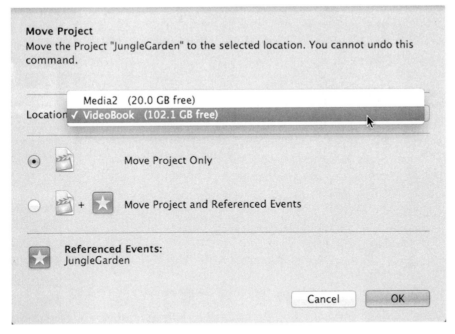

FIGURE 2.35

Move Project Dialog.

somewhat different dialog, as shown in Figure 2.36. Here you not only can select the drive to duplicate the project to, but also can select whether to duplicate just the project, the project with all its referenced events, or just the clips that are used in the project. This will not cut down (or "consolidate") the clips, but will include the entire clip for which any portion is used. You also have the option to include or not include render files. In some circumstances you'll want to use one, and in others not.

IMPORTANT NOTE

Saving

Final Cut Pro is not like other applications. There is no Save function, and there is no Save As because you don't need to save the document with a specific name. All FCP project files are always called *CurrentVersion.fcpproject*. They're in a folder with the name of the project. If you incorrectly named the project, you can rename it in the application, and the folder is renamed on your hard drive. There is a similar file for each event called *CurrentVersion.fcpevent*. There is a feature in FCP that not only saves the current project and current event, but also saves the previous version in a folder called *Old Versions*, which is inside the project folder.

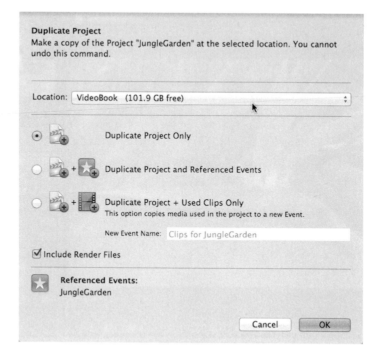

FIGURE 2.36

Duplicate Project.

You can create a new project at any time by

- Using **File>New Project**
- Pressing the keyboard shortcut **Command-N**
- Clicking the + button in the lower left of the Project Library
- Or, **Control**-clicking in the Project Library and selecting **New Project**

If you have multiple versions of a project, (we'll look at versioning in a later lesson), you can organize your project into folders by

- Using **File>New Folder**
- Pressing the keyboard shortcut **Shift-Command-N**
- Clicking the folder icon in the lower left of the Project Library
- Or, **Control**-clicking in the Project Library and selecting **New Folder**

I guess you're seeing a pattern here. There are usually two, three, or four ways to do most common functions in the application. Learn which ones work for you, which you're comfortable with, though you should be aware of the others because sometimes other ways might be more appropriate. For instance, if you're in the Timeline, neither the New Project button nor the shortcut menu will be available to you to make a new project, but **Command-N** will still work.

THE TIMELINE

Let's take a look at the Timeline window. To close the Project Library, you can either click the Film Reel button on the lower left of the window or press **Command-0** (zero). If you want to open a project, you can either double-click the project in the library, or select it and press the **Return** key.

Open the *JungleGarden* project to bring up the Timeline panel (see Figure 2.37).

In the center, running through the middle of the project panel, is the primary storyline with the clips on it. The clips contain both the video and audio and in this case have transitions, simple cross-dissolves between them, and a title connected to the first clip in the storyline.

At the top of the panel is the timecode displayed as a ruler, beginning at 0:00:00:00. The Timeline Ruler and the current time display will show time in whatever frame rate you're using in *Hours:Minutes:Seconds:Frames*. It can use either drop frame or non-drop-frame timecode. (See the sidebar "What Is Timecode?")

At the bottom of the Timeline window is another display that shows the duration of a selected item in the Timeline and the overall duration of the project (see Figure 2.38). If nothing is selected, the display shows the project duration and the format that the project is using (see Figure 2.39).

FIGURE 2.37

The Timeline.

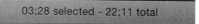

FIGURE 2.38

Item and Project Duration.

FIGURE 2.40

Buttons.

FIGURE 2.39

Project Duration and Format.

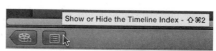

FIGURE 2.41

Lower-Left Buttons.

The name of the project you're working in appears in the upper-left corner of the Timeline panel. The buttons to the left of the name (see Figure 2.40) let you move between recently opened projects, just like the back and forward buttons in a web browser, and like a web browser, they use the same keyboard shortcuts **Command-[** for back and **Command-]** for forward.

There are two buttons in the lower left of the panel (see Figure 2.41). As in the Project Library, the Film Reel button will open and close the library. The Bullet List button next to it will open and close the Timeline Index (see Figure 2.42), which lists the items in your project and which we'll see in greater detail in later lessons.

In the upper-right corner are four buttons (see Figure 2.43). The first switches the skimmer off and on. You can also toggle it off and on with the shortcut **S**. When the button is blue, it's on; when gray, it's off. That's consistent throughout the applications. LEDs and checkboxes that are blue are on and active; gray or dimmed are off. The button with the waveform to the right switches audio skimming off and on. The shortcut is **Shift-S**. Note that when the video skimmer is off, audio skimming is off as well. Without the skimmer, dragging the playhead to scrub will produce no audio scrubbing.

The next button is soloing, which solos a single clip so that's all you hear. The shortcut is **Option-S**, and the last button will be familiar to legacy Final Cut Pro and Express users—similar in look, in the same position, with the same shortcut **N**—it is the button to toggle snapping on and off. With snapping on, dragging the skimmer or the playhead will make them want to magnetically stick to clips and edit points.

In the lower right of the Timeline is a slider that adjusts the scale of the panel content, allowing you to zoom in and out of the panel (see Figure 2.44). You can also do this with the keyboard shortcuts **Command- =** (think of it as +) to zoom into the Timeline and **Command- -** (that's minus) to zoom out of the Timeline.

Next to it is another toggle switch, the third in the FCP window, that is actually a pop-up which allows you to set the appearance of the clips in the Timeline. There are six display options from left to right (see Figures 2.45–50): audio waveform only; small filmstrip with large waveforms; medium filmstrip with medium waveform display; large filmstrip with a small waveform display; filmstrip only with no waveform; and clip names only.

Below the clip appearance buttons is another slider that lets you set the clip height from very small to very large; its use will be dictated by your monitor size, Timeline content, and eyesight.

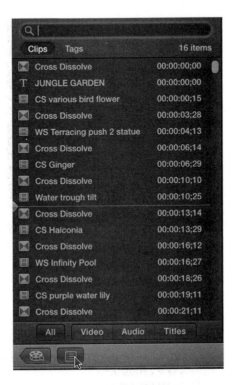

FIGURE 2.42

Timeline Index.

FIGURE 2.43

Upper-Right Buttons.

FIGURE 2.44

Zoom Slider.

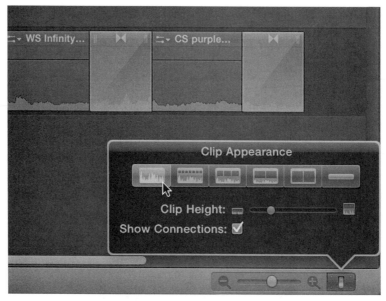

FIGURE 2.45

Audio Waveforms Only.

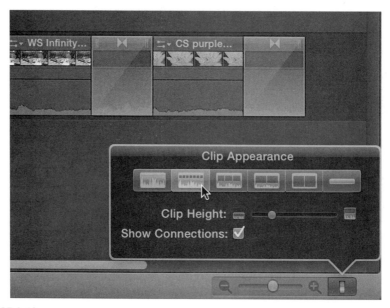

FIGURE 2.46

Small Filmstrip Large Waveform.

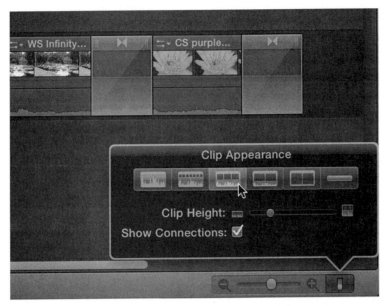

FIGURE 2.47

Medium Filmstrip and Waveform.

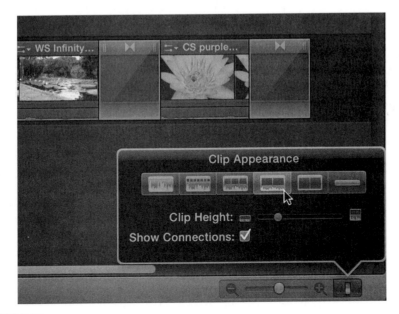

FIGURE 2.48

Large Filmstrip Small Waveform.

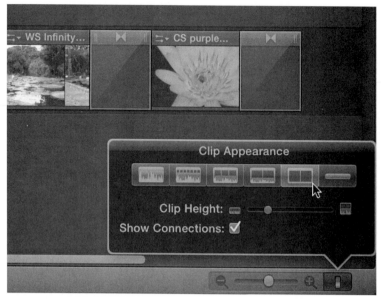

FIGURE 2.49

Filmstrip Only.

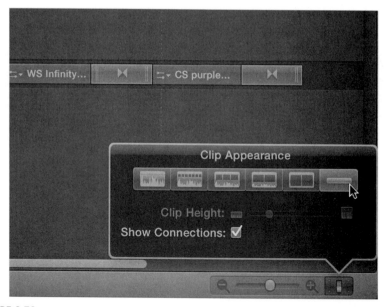

FIGURE 2.50

Clip Names Only.

TIP

Visual Stimulation

Many pro users find the "visual stim" of the thumbnails and frame views in the Timeline distracting, or perhaps they just want to more clearly see the edit points. Experiment with the choice and height setting you like best. The first choice gives you Audio Waveforms only, but you should be aware that with this choice you cannot double-click in the waveform region to separate audio from video to create an L or J cut, which we will look at in Lesson 7 on audio. If you drag the Clip Height slider all the way to the left with the choice second, no frames will be displayed. The fifth choice is video frames only, and the sixth is "slimline mode," which is convenient when many shots are stacked in a vertical edit. The second, third, and fourth choices give you the special ability to double-click the audio region of any clip to expand synced audio.

WHAT IS TIMECODE?

Timecode is a frame-counting system that is almost universal to video cameras. A number is assigned to every frame of video that is recorded. The numbers represent time, and on most consumer cameras, each clip begins at 00:00:00:00. On professional and some prosumer cameras, the start number can be set to anything you like, and the timecode can be continuous across continuous clips on the same memory card, or the timecode can be based on the time of day generated by the camera clock. A timecode number is assigned to every frame of video—25 or 50 frames per second in the European PAL system and 30 or 60 frames per second in the North American, Japanese, and Korean NTSC system.

This number can be a problem for NTSC because the true frame rate of all NSTC video isn't 30fps or 60fps, but 29.97fps or 59.94fps. Because of this, NTSC has created two ways of counting timecode called Drop Frame and Non-Drop Frame.

Non-Drop Frame (NDF) displays the numbers based on a simple 30fps frame rate. The problem with this is that when your timecode gets to the one-hour mark, one hour of real-world time has already come and gone. The one hour of clock time finished almost four seconds earlier, so your program is running too long.

Drop Frame (DF) uses a complex method of counting that compensates for the difference between 29.97fps and 30fps. No actual frames of video are dropped. DF drops two frames per minute in its count, except every tenth minute. This means that at the one-minute mark, your DF video will go from 59;29 to 1:00;02. There is no 1:00;00 or 1:00;01. Notice the semicolons. The convention is to write DF timecode with semicolons, or at least one semicolon, but NDF is written only with colons.

TIP

Help

If you have problems with Final Cut Pro, help is available in a number of places, including the Apple Support Communities Final Cut Pro discussion forum, which you can link to from https://discussions.apple.com/index.jspa. You can also get help at the Los Angeles Final Cut Pro User Group website forum at www.lafcpug.org/phorum/index.php or my website at www.fcpxbook.com.

QUICK EDIT

We are going to look at the application in some detail over the course of this book, but you'll probably want to get started right away, so I want to give you a quick overview of editing in FCP. There are many ways to edit in FCP, using many tools and techniques, but this is the simplest and probably the one most familiar to iMovie users. Let's begin by making a new project:

1. To make a new project, press **Command-N**.
2. Name the project *QuickEdit*. Because you didn't select any specific location, the project will probably be saved in the FCP1 sparse image, which is fine as this will be a short project.
3. Click the *Favorites* keyword collection inside the *Collections* folder in the *JungleGarden* event.
4. Make sure the Event Browser is in List view and select the second shot *CS Heliconia* that's in the list.
5. Skim the shot to around the 10:00 point, just after a camera bobble.
6. If you've clicked the clip to select it and it has a yellow box around it, press the **I** key to mark an In point. If the whole clip is not selected, you can mouse down and drag across a clip to create a selection that's about five seconds long. If you marked an In point, skim forward about five seconds and press **O** to mark the end of the selection.
7. Move the pointer over the selection area, and the Hand tool automatically appears (see Figure 2.51). Drag the selection and drop it anywhere in the Timeline. It will move to the beginning of the project.
8. Let's add another shot. Drag a selection through *CS Yellow water lily* in the browser.
9. Drag it to the Timeline and drop it. It will be magnetically attracted to the end of the first clip. These are called *append* edits, as they are always appended onto the end of the project.
10. Drag a selection in *MS Pineapple* toward the end of the shot after the zoom finishes.

FIGURE 2.51

Hand Tool and Selection.

11. This time, drag the clip into the Timeline and put it right between the two other clips so they move out of the way, as in Figure 2.52. This is an Insert edit.

12. Click in the Timeline to make it active or press **Command-2**.

13. Press **Shift-Z**, which is Zoom to fit and one of my favorite shortcuts. It will fit the contents of the panel into the available space.

14. Select all the clips in the Timeline with **Command-A** and press **Command-T** to apply the default cross-dissolve transition to all the clips.

15. To add a title, click the **T** button in the button bar, and a browser with the various titling options appears (see Figure 2.53). Scroll down till you find the Fade title.

16. Select the Fade title and drag it above the first clip in the Timeline so that it appears over the clip and becomes connected to it (see Figure 2.54).

17. Select the title in the Timeline, and it appears in the Viewer. Double-click the text in the Viewer and type in *Jungle Garden*. Press the **Escape** key.

TIP

Picking Where to Save Your Project

If you want to save your project in a specific hard drive, go to the Project Library and select the drive or partition before you create the new project.

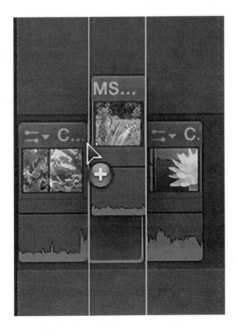

FIGURE 2.52

The Insert Edit.

FIGURE 2.53

The Text Browser.

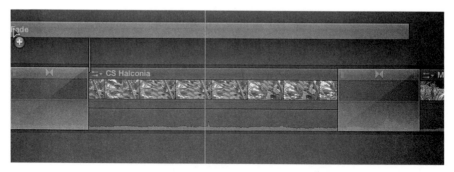

FIGURE 2.54

Title Over Clip.

That's it. You're editing in Final Cut Pro. If you're an iMovie user, you've worked pretty much exactly the same way you worked in iMovie. If you have to quickly lay out some shots and upload them to the Web, you're ready to go. There's a great deal more to the application, of course, as we'll see in the chapters ahead, but you can get started playing with it and having fun.

SUMMARY

You should now be familiar with the interface and the some of the terminology used in Final Cut Pro. The Quick Edit got you started on a basic way of working with the application. I'm sure at this point that you have many questions, and I hope to answer as many of them as possible over the course of these lessons so that you can work smoothly and efficiently with the application. Spend some time clicking around in the Final Cut Pro window. You can't hurt anything. And remember to try **Control**-clicking to bring up shortcut menus. In the next lesson, you'll learn how to set your preferences for the application and how to get media into your system so you can start making your own projects.

Preferences and Importing

3

LESSON OUTLINE

Digital video editing is divided into three phases:

1. Getting your material into the computer
2. Editing it (which is the fun part!)
3. Getting it back out of your computer

This lesson is about number 1. Before you import your media, you should set up your application correctly for the way you like to work and also so that it's set up for importing. In Final Cut Pro, as in most video editing programs, that means setting up your preferences.

After setting preferences, we'll proceed to importing media. For this lesson you may need to download some materials from my website. If you don't have any AVCHD or file-based media to work with, you can go to www.fcpxbook.com. On the website there are a number of ZIP files, each of which contains a sparse disk image that we will work with. For this lesson we need the ZIP file called *NO_NAME.sparseimage.zip*. This is about 300MB in size, so it may take some time to download. The sparse image itself is even a little larger when it's unzipped. We'll use this media when we look at importing from a file-based camera.

USER PREFERENCES

For Final Cut Express users, who had two separate preferences settings with a total of 9 separate panels, or legacy FCP users, who had three preferences settings with no fewer than 17 panels, the simplification of the new FCP preferences must be a blessing, though, perhaps also a bit of a worry. It might be nicer if there were more user preferences, and perhaps more will be added as time goes on.

Start by double-clicking the Final Cut Pro icon in the *Applications* folder or single-clicking the icon in the Dock or using Launchpad in Lion, whichever way you prefer to open the application. Double-clicking a specific *CurrentVersion.fcpproject* in a project folder is of no benefit in FCP, as the application always opens in the last-opened state, and all available projects are always seen in the Project Library.

To access the preferences, go to the **Final Cut Pro** menu and select **Preferences (Command-,)**. When you open preferences, you will see the window in Figure 3.1, which first opens on the **Editing** tab, the first of only three. If you've already opened preferences, it will open to the last-used tab. We'll start with the Editing tab.

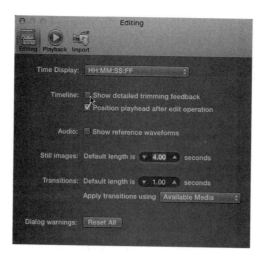

FIGURE 3.1

Final Cut Pro preferences.

Editing

The first pop-up for **Time Display** defaults to HH:MM:SS:FF (Hours:Minutes: Seconds:Frames), which is fine for most uses. You can also change it to include subframe amounts, which can be useful when you're doing precise audio editing (see Figure 3.2). Although you can audio edit in subframe values, video is frame based only and plays in exact, specific frame values. At 24fps, humans can't see that they are watching separate, individual pictures. This is called *persistence of perception*, and is what makes film and television possible. The fact is that you cannot see and hear the difference in sync between one frame of video and audio. But you have indicators that will show subframe audio timing if you need it. In this pop-up you can also select to view your media as a frame count, which is commonly used by animators and in film work. You can also view in whole seconds. I recommend leaving it at the default.

In the **Timeline** options, the first item is checked off by default and really should be checked on and left on. You really do want to see detailed trimming feedback when you're editing.

There is a checkbox for **Position playhead after edit operation**. This will move the playhead to the end of the edit you just performed, which is the default behavior in FCE and probably what most users would expect. You can switch it off if you prefer.

In **Audio**, Show Reference Waveform can be useful, but I'd leave it off for now, as it clutters the Timeline a little. When we get to audio editing, you can turn it on.

For **Still Images**, you can set the import duration to whatever you prefer. Four seconds is the default. Notice you cannot set seconds and frames, but you can set seconds and a decimal value. Because there are now so many frame rates in use, setting an import value of four seconds and 30 frames makes no sense for PAL-frame rates of 25fps but makes sense for video if your frame rate is 60fps. Hence, seconds and decimal values in the value box.

The **Transition** default is one second, which seems fine for me, but if you prefer something else, change it. Again, the value is set in seconds and decimals rather than seconds and frames. If you'd like less than a second, you can highlight the box and type in decimal parts of a second, such as 0.1 for one-tenth of a second.

FIGURE 3.2

Time Display Preferences.

FIGURE 3.3

Transition Preferences.

The next pop-up defaults to **Available Media** (see Figure 3.3), which is what professional editors have always used. This behavior is what Final Cut Express editors are used to, and it does not change the project duration, nor is anything shifted in the Timeline. It does require you to provide extra media for the transition called *handles*, which we'll look at in Lesson 8 on transitions. On the other hand, **Full Overlap** is what iMovie users are comfortable with. Full Overlap means that every one-second transition you apply will shorten your project by one second and shift everything in the Timeline. This should not be the preferred option.

And finally there is the **Dialog warning** button, which you can click to reset the warnings at any time to their default state—a useful little feature.

Playback

The **Playback** tab of preferences allows you to change how the application behaves (see Figure 3.4). The first setting for **Rendering** allows you to turn background rendering off and to set how quickly it kicks in. The lowest setting is one second. You can actually set it to zero, but it will still wait at least a second before it starts, and if you do set it to zero, it will reset to five seconds the next time you open the application. The five-second delay doesn't bother me, and I'm leery of setting it too low. When you get into heavy compositing and effects work, you actually may prefer to switch off automatic rendering and initiate rendering whenever you need it. You can force rendering with the **Modify>Render Selection (Control-R)** function, which will render a selected item, or to render everything, use **Modify>Render All (Control-Shift-R)**.

WHAT IS RENDERING?

A project is only an instruction set to your computer on how to play back your media. When you play back your project, you're playing back the media that's on your hard drive, the movies you shot and imported from your camera. As you start adding effects, transitions, titles, special effects, color correction, it's as if you're creating a new clip—the media for these doesn't exist anywhere. If your computer is fast enough to process these effects in realtime, it will play back the material without problem; if it isn't, it will try to process the material but will skip frames during playback, giving the video a stuttering effect during playback. In either case the media for that piece of your movie doesn't actually exist on your computer. What rendering does is create a small QuickTime file that covers that area of your project that has the effects, allowing them to play back in realtime. The render files for each project are saved in a folder inside your project folder. In this case that render folder will be inside the sparse image, and the render files will be created on that mounted disk image.

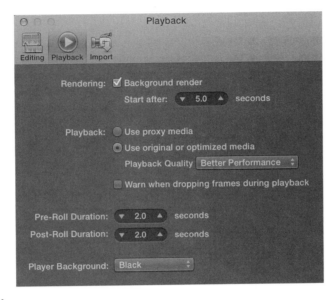

FIGURE 3.4

Playback Preferences.

> **NOTE**
>
> Background Rendering
>
> Apple marketing calls what FCP does background rendering, but the truth is that it's not. It's *seamless* rendering or *automatic* rendering. It renders when it has time, when you're not doing anything. It is initiated after a chosen delay and will pause when you start to do something in the application and resume when it can. It does not actually render while you're doing other things in the application. However, you do not have to do anything to initiate it or to pause or stop it once it's begun.

The **Playback** section itself lets you select whether to play back using proxy media or original or optimized media. You can switch this at any time, and you need to switch it if you want to work with proxy media and then switch it to play back original/optimized media when you're exporting. It's important to understand that currently, if you are working with proxy media and you want to output high-quality material, you must go into this Preference setting and switch on original/optimized playback. Otherwise, you will output a low-resolution, proxy-quality file. We'll look more at these options in the section on importing in this lesson.

Playback quality defaults to **Better Performance**, which is probably a good choice while you're editing. If you want to show someone your project in full screen, playing back from the Project Library or Timeline, it might be worthwhile switching to **High Quality** playback.

The line for dropped frames is checked off by default, which may seem strange to Final Cut Express users; in FCE, the default state is to have it on. In FCP, because so much is done in realtime, even when things are unrendered, leaving this setting unchecked is essentially the same as using the Unlimited RT option that was available in Express and legacy FCP. You could play through material that needed rendering with the occasional dropped frame even when rendering was required. This is probably why the FCP render bar is the same orange color as the unlimited RT render bar that we saw in legacy applications—a change from iMovie '11 where the Event Library simply indicates footage used in a project.

Pre-roll and post-roll duration can be set to whatever you like. This setting controls how far back in the Timeline playback begins and how far ahead past the current position the playhead travels when you use the Play Around function, which is useful for reviewing cut-points. I like two seconds before and two seconds after, but you can pick whatever you prefer, perhaps the traditional five seconds of pre-roll and two seconds of post-roll.

Player Background can be set to Black, White, or Checkerboard. This means the Viewer background display. The default is black, but the background actually is transparent. As a result, text over an empty background is text on transparency, so you might prefer to use checkerboard as the background to show that, although checkerboard during playback does look strange.

TIP

Improving Playback

If you're having difficulty playing back your media because your system is underpowered and the format, such as AVCHD, is too processor intensive for your computer, you can help playback considerably by during off the audio waveforms in the Event Browser or in the Timeline. Select a Timeline display that shows only video filmstrips, and the processor will not have to rebuild the waveform as it plays the project. Removing this extra load can be a big benefit in improving video playback performance.

Import

The **Import** tab of preferences is the most important of the three preferences (see Figure 3.5). This tab sets your default importing behavior and is also used as your setting whenever you drag and drop to import something into your Event Library.

By default, the first two items are checked on, both of which seem like a good idea. If your media is already on a drive and you don't want to move it, or you have to connect to media that's on a network, you can uncheck the first box and FCP will simply create an alias that points to the media in whatever location it's in. You should not move, rename, or in any way alter the media at the Finder level, as FCP will lose

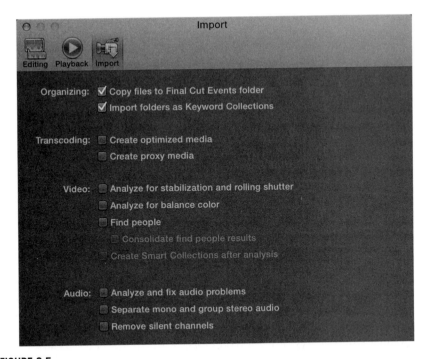

FIGURE 3.5

Import Preferences.

the connection to it. Checking the box for **Import folders as Keyword Collections** will be discussed later.

The two checkboxes for **Transcoding** are important. There has been a lot of discussion about what is the best practice for handling your media. The general recommendation is that when you're importing compressed formats such as H.264, which is used in AVCHD and in DSLRs, it's best to optimize the media. This produces a QuickTime file using the ProRes codec, a visually lossless, high-resolution, high-data-rate format that produces very large files. This is will improve playback performance but requires very fast 7200rpm hard drives, using FireWire 800 or faster. FireWire 400 may work, but it will be marginal performance. When H.264 is imported using optimization, FCP imports BOTH the original media files and separate ProRes files, though it uses ProRes for playback, if available, not the original files. If your computer is fast enough and has a good, fast graphics card, I suggest working with H.264 is not a problem and should be used, saving yourself a great deal of processing time and hard drive space. Some formats such as DV and HDV cannot be optimized in FCP, as the application handles them in their native codec. At any time in the editing process, you can transcode specific files or even all your files to standard ProRes 422 or ProRes proxy files.

Proxy files can be created if your computer is an older model and cannot process H.264 efficiently, and you do not want to expend the drive space required for ProRes files. These are very small, quarter resolution QuickTime files created using the ProRes Proxy codec. You can edit using the proxy files by switching on proxy playback in the Playback tab of preferences, and then when you're ready to output, switch back to the original media, which FCP does instantly and seamlessly. If you're working with proxy files, it would be a good idea to switch to original media before you do effects work and color correction so that any such effects work is rendered in the higher-quality codec.

An important new feature in FCP is its ability to analyze and correct media on import. There are separate sections for video and for audio analysis.

The first two video analysis options simply check your media for stabilization or rolling shutter problems and for color balance issues. They do not apply any correction. You have to manually activate these options for the clips in the project that require it, but the analysis can be done on importing. These analyses can be quite slow, requiring a lot of background processing. These can be run at any time later, however, after you have imported your media. To activate analysis in the Event Browser, select the clip or clips you want to have analyzed and **Control**-click to choose **Analyze and Fix**. This brings up a dialog box that is identical to the video and audio sections we see here in the importing preferences. You can also activate this from the **Modify** menu. Unless you have significant color balance and obvious stabilization and rolling shutter problems (a commonly seen "weaving" effect from DSLR cameras during movement such as pans), I recommend leaving this checked off on import. Generally, color balance is best done manually. I find the automated color balance a little weak and leaning a little too much to blue for my liking, but try it; it might suit you.

NOTE

Background Processes

One gotcha you should be aware of is that there are a number of functions in the browser that cannot be performed while analysis and transcoding are going on. For instance, you cannot move a file to the trash, rename a file in the browser, or move or duplicate an event; there are a number of other restrictions. You also cannot quit the application during these background processes without force-quitting the application. These are even more reasons to minimize the processing time during importing.

Find People? That's right! FCP will analyze the media for content, looking for the shot type—close shot, medium shot, wide shot—and for the number of people in the shot—one person, two people, and groups of people. All of this analysis will further slow down importing. The people analysis also activates shot-type analysis. The latter works pretty well, and if you find that useful, you might wish to activate this feature. The people analysis for numbers of people I find to be pretty poor, often producing bad results or apparently randomly breaking up clips when it seems to lose sight of people in the frame. The Consolidate option instructs the application to use the number of people that is predominant in the shot as the basis for its labeling. Smart Collections to store these can automatically be created based on these analyses. Just use the last checkbox in the Video section.

The Audio section is quite different and, unlike with the video options, I usually check these all on. The audio analysis is quite a bit faster, and it does more than analyze: It actually applies any fixes it finds necessary, such as background noise reduction, hum removal, and a loudness adjustment. Separating mono and grouping stereo tracks seems like a good idea to me, as does removing empty audio channels, which often happens with cameras that shoot four or eight tracks of audio. Any time you shoot using a separate mic that records in mono, you are likely to have an empty audio channel.

Preferences Files

If you have problems with FCP, one of the first remedies anyone will suggest is to trash your preferences file. If there is a problem with your system, it's often your FCP preferences that are corrupt. Follow these steps to delete them:

1. Make sure the application is closed.
2. Go to your *Home* folder. Press **Shift-Command-H** from the Finder.
3. The preferences files are in the Library. Unfortunately, in Lion 10.7, Apple has thought it best to hide the Library. To open it, hold the **Option** key and use the **Go** menu in the Finder to select it.
4. Inside the *Library* folder, go to the *Preferences* folder.
5. Find the *com.apple.FinalCut* files. There may be several of them. Select and delete them by dragging them to the Trash or selecting them and pressing **Command-Delete**.

This is the manual way to perform this operation. A better way would be to download Preference Manager from Digital Rebellion at www.digitalrebellion.com/prefman. This is a free component of Digital Rebellion's Pro Maintenance Tools. I urge you to get the full package, as it has many useful tools, and Jon Chappell at Digital Rebellion makes excellent products for Final Cut Pro users.

CREATING A NEW EVENT

Now that we have set up our preferences and everything is ready, we can start bringing our own material into the application. The process is called *importing*. Whether we're importing from iMovie or importing from a tape-based or file-based camera, camera archive, or our hard drive, it's all importing. All of these processes are essentially data transfers, whether from a hard drive, a memory card, or digital tape. Whenever you import media, it gets put into an event. So let's begin by creating a new event. For every production you should have at least one separate event. Often one is enough, but if you wish, you could create a separate event for each card you shoot (though this can complicate search functions) or any other organizational routine you prefer. You can create a folder in the Event Library to group events if you wish. I usually use just one event for one production. That keeps all the original media in one location and makes a clip search easier, as I don't have to select multiple events before I can begin the search.

1. In the Event Library select the media drive on which you want to create your new event and choose **File>New Event** or use **Option-N** or **Control**-click and select **New Event**.

2. The event is created, and its name is highlighted, ready to be renamed as in Figure 3.6. Let's call the event *Importing*. This will actually change the folder name on your hard drive. The event is saved in a subfolder in the *Final Cut Events* folder on the designated drive.

A new event without any media gives you two helpful buttons to help you start importing your media (see Figure 3.7). The event will hold any media you import into it—video clips, audio files, and still images. Every event folder on your drive will contain a document called *CurrentVersion.fcpevent*, which defines the state and content of that event. As you import the media, other folders get created in the hard drive's event folder, such as in Figure 3.8.

NOTE

Shared Area Network

Whenever either the Event Library or the Project Library is the active panel, you have the option in the **File** menu to choose **Add SAN Location**. This allows you to designate a shared area network that's mounted on your computer as a location to create your events and projects. This needs to be something like an Xsan connected by Fiber Channel. It means these items can be accessed and used by others on the same SAN. Only one user can access them at a time, however. Once you add it to your interface, it's locked out to others. You have to use **Remove SAN Location** to release it to other users.

FIGURE 3.6

New Event in Library.

FIGURE 3.7

Event Import Buttons.

| Analysis Files | CurrentVersion.fcpevent | Original Media | Render Files | Transcoded Media |

FIGURE 3.8

Event Folder Content.

IMPORTING

Let's go through the different ways we can import content into our event. If the files are already on your computer or a connected hard drive, simply use **File>Import>Files** or **Shift-Command-I**. This will bring up the dialog in Figure 3.9. Here, you have the option of either importing to the selected event or creating a new event and designating the hard drive for the event. At the bottom of the dialog box are the default importing options as you created them in your preferences, and you have the opportunity to change them, depending on your media. To do the import, just navigate to where the files are located. Next, select the items—**Shift**-clicking to select contiguous items or **Command**-clicking to select non-contiguous items. Then just click the **Import** button or press the **Return** key.

This is also the way you import from DSLR (Digital Single Lens Reflex) cameras that shoot in QuickTime format and create .mov files. You connect the camera or card reader and make sure it's mounted on your desktop. If you have backed up or archived the media from your camera, you can open the backup folder or mount the camera archive and navigate to the folder, usually called the *DCIM* folder, that contains your video files. Select them and import the clips.

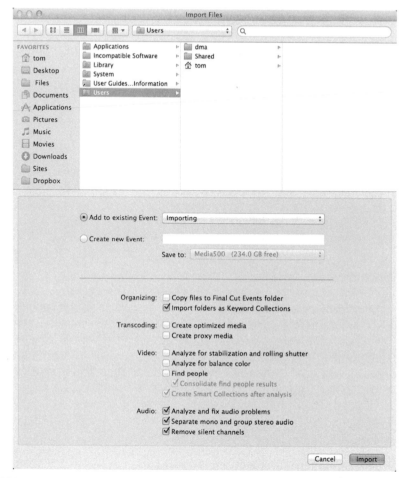

FIGURE 3.9

Import Dialog.

TIP

Mixed Media on a Card

Often on your DSLR card, you will have a mixture of stills and video randomly shot. Here's a good tip from Thomas Emmerich. There is a little pop-up to the left of Users in Figure 3.9 that allows you to select how the material is ordered in the Import dialog. Normally, it orders by Name, but you can set it to order by Kind, which will separate the stills from the video. This technique can be useful if you just want to import your video.

You can also import a file by dragging and dropping from a Finder window into the Events Library; directly into a Keyword Collection, which will apply the keyword to the clips; or even directly into a project. The media will be imported based on your import preferences. If copy is unchecked in the preferences, aliases will be created pointing to the media. If copy is checked on, the files will be copied into the *Original Media* folder in the event folder. If you drag directly to the Timeline panel, the media is also placed in your designated default event for the project. If you drag audio from a CD to import, which you can do, the audio will follow your preferences, so be very careful. If you have the copy function switched off, the imported track will simply be an alias pointing to the track on the disc. It's always a good idea to convert the media from a CD before importing it. We'll see how to do this later in the lesson.

> **NOTE**
>
> Changing the Default Event
>
> If you already have a project whose default event is on the internal drive or a media drive and you want to change the default event to something else, use **Command-J** to open the project properties. Click the wrench icon at the bottom right of the Inspector and change the default event for the project.

IMPORT FROM CAMERA

Importing from a camera, either file-based or tape-based, is really simple. Let's look at importing from a file-based camera first, as that's probably what most users are now working with. This is very similar to importing from a camera in iMovie and analogous to the log and transfer function in Final Cut Express.

1. Connect the camera in Playback or Computer mode, or insert the memory card into a card reader or the slot on your computer if there is one.
2. To initiate import, either click the Camera button on the far left of the Toolbar (see Figure 3.10) or use **File>Import from Camera** or simply press **Command-I**. Often the camera is detected when it's connected, and the application automatically launches the camera import function.
3. With the Camera Import window open with no camera detected, you will most likely see yourself from your computer's built-in FaceTime camera. Unless you want to record yourself (which you can do by the way), by clicking the Import button, you should select the memory card that should be listed under CAMERAS on the left side of the window.

FIGURE 3.10

Camera Import Button.

> **NOTE**
>
> Camera Mode
>
> It's important that you put your camera in Media or Playback mode rather than in Camera mode when you connect it. If your file-based camera is not in Playback mode, it may not be recognized by the application.

Creating Camera Archives

Most video media is now recorded as files on an SDHC card, a CF card, a P2 card, or a hard drive. All of these are camera media, designed to be reused, not stored. Before you import your media, back up the media from the card or hard drive. The first step you should always take with file-based media is to archive it. The tiny memory card is the only recording of your production or of a precious, never-to-be-repeated moment in life. Make another copy before you do anything else. It's really simple to do in FCP. When the application detects the media, it opens the Import window and populates it with the card's contents.

1. Select the card, which is usually called *NO NAME*, in the camera list as in Figure 3.11.
2. At the lower left of the Camera Import window, click the **Create Archive** button as in Figure 3.12.
3. A dialog window drops down that lets you specify which drive you want to archive the media onto. Select the drive.
4. Name the archive something useful such as the event and the date or the name of the production followed by a number.

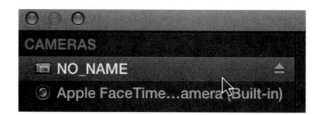

FIGURE 3.11

Memory Card in Camera List.

FIGURE 3.12

Create Archive Button.

As with events and projects, you don't get to decide where the folder it creates, *Final Cut Camera Archives*, is placed. As with the other folders, this one is also either in your home *Movies* folder or on the top level of the selected drive.

The drive used for archiving should not be your media drive. It should be a separate drive—not necessarily a fast drive—but it should be large. This becomes your archive storage. Keep it safe.

You can also archive your media manually. There are two ways to do this.

The first way creates an archive folder into which you drag your card contents:

1. Create a new folder on your hard drive and name it something useful.
2. Select the ENTIRE contents of the memory card, including ALL its subfolders. Do NOT drill down into the folder structure to find the video. FCP must have the whole folder structure.
3. Drag the entire contents of the card into the archive folder.

The second way creates a disk image:

1. From your *Utilities* folder in *Applications*, open the Disk Utility.
2. On the left select the mounted disk called *NO NAME*, or whatever it's called.
3. Go to **File>New>Disk Image from "NO NAME."**
4. Give the disk image a more useful and appropriate name and save it wherever you like on your archive drive.

One of these two methods should always be used to archive media from a DSLR memory card. This is essentially the same process that FCP is doing for you, except it always puts the archive in a specific place and gives the archive a nifty film can with keyhole icon. It also cannot be as simply opened as the Disk Utility–created archive, which can be double–clicked to mount the disk and its contents.

File-based Camera

Importing from either a camera or a camera archive is exactly the same. The only difference is that the camera archive will not be immediately recognized by the application, and you have to click the **Open Archive** button in the lower left. Once the archive is mounted, you can select it in the panel on the left of the Camera Import window, as in Figure 3.13, which is also in the color section. Let's see how you use this window.

If you don't have file-based media of your own and you want to follow along, you'll need to download and unzip the *NO_NAME.sparseimage.zip* from my website. Go to www.fcpxbook.com to get the media. Then unzip the file by double-clicking on it, and then double-click the sparse image to mount it.

1. In the FCP Camera Import window, select the memory card under Cameras on the left, which will fill the Import window with your clips.
2. You can click the **Import All** button in the lower right if you wish.

FIGURE 3.13

Camera Import Window.

3. You can also select clips by clicking on them and then using the Import button, which has now changed to **Import Selected**. You can also simply press the **Return** key.

4. You can play your clips with the buttons in the Viewer or with the usual keys— **spacebar** as well as **JKL**—and mark selections with the **I** and **O** keys to select a range in a clip. You can mark a selection on only one clip. Trying to mark a selection on another clip will cause you to lose the first selection.

5. You can also drag a selection on the clip with the pointer.

6. Once you've clicked to import one selection, you can immediately make another selection to start importing it.

Each time you click the Import button, you will get the Import dialog as in Figure 3.14. This is very similar to the standard Import dialog. It allows you import to the selected event or to create a new event and designate its location. You have all the standard importing and analysis options except you do not have options to disable Copy events. You have to copy the media from the memory card or archive. You cannot link to it. The media should always be copied to your separate big, fast media drive. This is especially important if you are optimizing your media.

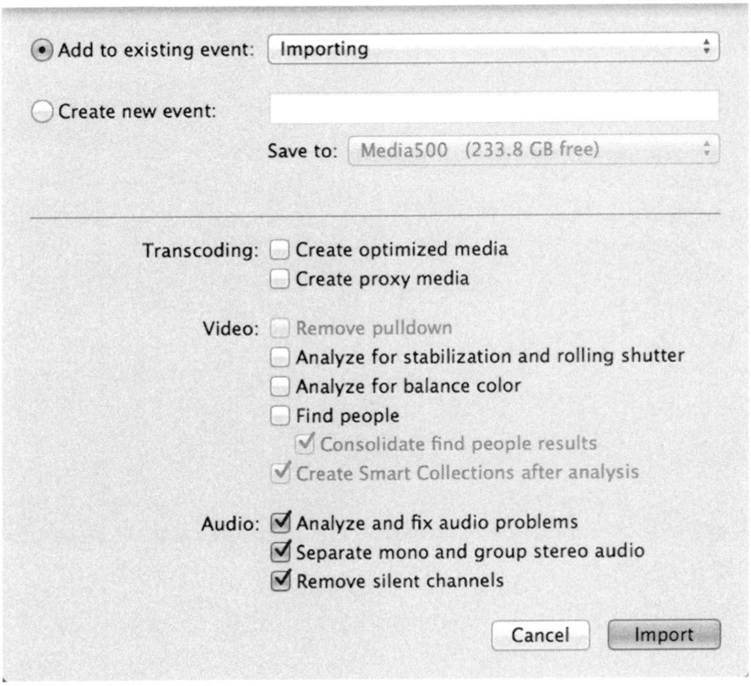

FIGURE 3.14

Camera Import Dialog.

The archive drive and the media drive should be separate so that you have two copies of your media.

After you've imported a clip, an orange bar will appear on the clip or the portion of the clip in the Import window. Notice the checkbox in the lower left under the clips that will let you hide previously imported clips. Also notice on the right the slider and light switch, which work exactly like the slider and light switch in the Event Browser when it's in filmstrip view. The slider controls how many icons you see as a filmstrip in the Camera Import window. It defaults to All so that each icon is a separate clip. The light switch pop-up, in addition to adjusting the clip height, lets you turn on audio waveforms in the window. This capability can be very useful to see where there is sound and give you a first indication of their levels (see Figure 3.15).

TIP

Quickly Importing

Once you have the Import dialog set up, you can make a selection in the Camera Import window and quickly import by pressing **Return** to import and then pressing **Return** again to accept the dialog.

FIGURE 3.15

Clips with Audio Waveforms.

Clips from AVCHD media, such as we're using here, are usually numbered sequentially by the camera. When FCP imports the clips, the filenames are changed to the date/time stamp indicating when the material was shot or at least the time the camera clock was set to when the material was shot.

Importing from an iPhone, iPad, or iPod Touch is the same as importing from a camera. You simply connect the device using the USB cable and the FCP Import from Camera functions.

> **NOTE**
>
> Spanned Media
>
> *Spanned clips* are clips that bridge across multiple memory cards. To import them, you need to make a separate camera archive of each card that contains the spanned media. Make sure all the camera archives are mounted and available, select them, and click **Import All**. If only some portions of the spanned clips are available, the camera archives can still be imported, but they will remain as separate clips and not be joined together when they are imported into the FCP Event Library.

Reimporting Files

Sometimes clips might get disconnected while you're editing, and it may become necessary to reconnect them. You can do this from the Project Library. Select the project in the library with the missing clips, which should display a little exclamation warning icon next to the filmstrip (see Figure 3.16). Then use **File>Import>Reimport from**

FIGURE 3.16

Media Offline Icon.

Camera/Archive. Simply click the **Continue** button in the dialog that appears, and if the archive is mounted and available, the application will immediately reimport the clips into the appropriate event and reconnect the clips in the project. This function works only for camera media and camera archives, which makes it even more imperative that you back up and archive your camera originals and save them on a secure drive, separate from your media drive. It is the only way to re-access media that has been misplaced or moved and appears with the dreaded red offline media icon in the Viewer.

NOTE

Reconnect

There is currently no function for reconnecting media that becomes offline, such as in Final Cut Express and legacy versions of Final Cut Pro. You cannot reconnect to any file you want, nor can you reconnect to a file if the name has been changed, though the application will detect files with the original name and specification in the same location.

Importing from Tape-based Cameras

FCP can import from tape via FireWire, so only cameras or decks that have FireWire output and can be connected to your computer via FireWire will work. For most people, this means importing either from DV or HDV tape. This is a very simplified import function, very much as it is in iMovie and similar to the Capture Now function in Final Cut Express.

When you connect the camera in Playback mode, the application will usually detect it and immediately open the Camera Import window (see Figure 3.17). Here, you get a viewer with controls to play and rewind your tape. Unlike in legacy Final Cut Pro, you have no ability to make selections by marking In and Out points, logging your material to make selections, or batch capturing from your logged list. To import portions of the tape, you simply have to do it manually. Queue the section and click the **Import** button. Clicking Import will bring up the dialog in Figure 3.18. Notice that for DV and HDV material, there is no option to optimize media because it's already optimized. You do get the option to create proxy files, though, which will be done after the data is imported.

To stop importing, you can either click the **Stop Import** button or press the **Escape** key.

When you import from tape, the application automatically splits the material into clips when it detects a shot change caused by a break in the date/time stamp.

FIGURE 3.17

Tape Capture Window.

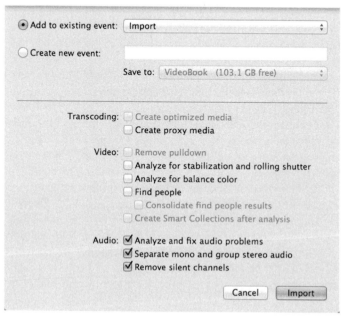

FIGURE 3.18

Tape Import Dialog.

It's important for DV tape cameras that the internal clock be set, not necessarily to the right time, but that it be set to some time. As in AVCHD importing, the clips are named based on the time stamp the application receives from the camera clock.

You can also archive a tape, though most people use the tape itself as the archive. However, a camera archive can be used to reconnect media if necessary, whereas a tape cannot. To archive a tape, click the **Create Archive** button in the lower-left corner of the window. The application will seize control of the tape, rewind it to the start, and starting playing it back to import the clips, which all get put in the locked camera archive. DV and HDV files are around 13GB per hour of media, so these will be much larger archives than most AVCHD camera archives.

Importing from iPhoto and Aperture

To access stills from iPhoto or Aperture, click the camera icon in the Toolbar (see Figure 3.19). You can also use **Window>Media Browser>Photos**. Here, you can access your libraries, events, albums, and projects. To bring the stills into your Final Cut project, simply drag the image into an Event in the library or directly into the Timeline. Still images that are dragged into the application are always copied into your event folder rather than aliased.

FIGURE 3.19

Importing Stills.

When still images are placed in a project, they will conform to fit the window by default—scaled down if the image is larger than the screen, scaled up if the image is smaller than your frame size. This can be controlled in the Inspector, as we'll see in Lesson 12 on animation.

> **NOTE**
>
> Adding Stills
>
> Be careful about adding stills to the project first. The first item in a project usually sets the project's properties. If you add a still image first, FCP will select 23.98 (actually 23.976) as the frame rate. Always put a video clip into the project to allow the project to set the frame size and frame rate. Once there are clips in the project, the frame rate can no longer be changed, and you can add stills without problem. You can delete the clip once you've put it in the project, but do so by actively deleting, not by simply undoing, because that will undo setting the project properties.

You can import both video and stills from iPhoto or from Aperture. However, I strongly advise that you import your videos directly into Final Cut and your stills into the photo applications. You should not use the photo applications to import video. iPhoto and Aperture videos may not appear in the Photos browser in FCP. If you have video in these applications that you have to import, you can open both applications and then drag and drop the videos from iPhoto or Aperture to an event in FCP. In this instance, if copying is turned off in the import preference, only an alias is created that points to your iPhoto or Aperture libraries. Again, I strongly recommend against doing this.

You can also import stills from anywhere on your hard drive using the import files function (**Shift-Command-I**). This method will not copy the file if the *copy files to Final Cut Events folder* option is off.

Changes made to images in Photoshop or Gimp or any image editing application will be recognized in FCP if the changes are made while the application is open. This sometimes works even if the application is closed, but it's not reliable. Any changes will be updated in the FCP event and Timeline automatically. The only thing that will not be tolerated is if the file is renamed. It is no longer necessary to use an Open in Editor function, as in Final Cut Express.

Importing from iMovie

You can also import from a project or your event library from iMovie using **File>Import>iMovie Project** or **iMovie Event Library**. You should be warned, though, that some content in an iMovie project will not import. For instance, iMovie trailers do not import the graphics or the music. Even converting a trailer to an iMovie project in iMovie will not bring this media into FCP. The only way to get it into FCP is to export to QuickTime from iMovie.

Similarly, iMovie-generated content like the animated globe does not import into FCP. However, many titles do import and can be edited in FCP Title Inspector.

iMovie projects and events are always imported into your home *Movies* folder. If you need to use the media, you should use the **File>Move Project** and **File>Move Event** functions to move these to your media drive.

Some media that iMovie imports does not transfer correctly to FCP either. For instance, iMovie captures DV tape using the native DV Stream format. This is a multiplexed format; that is, the audio is embedded and is part of the video data stream. When FCP imports DV, the file is converted to QuickTime, and the audio is stripped out and added to separate tracks. The iMovie DV footage comes in simply as DV Stream, and the audio cannot be heard or used in FCP. You could either export the video from iMovie as QuickTime or reimport the media from tape into FCP. Furthermore, iMovie significantly degrades DV and HDV material used in an iMovie project, so it's better to use FCP to import this media.

Importing from XML

FCP uses a new form of XML (Extensible Markup Language) to transfer information between applications. This allows you to export a small file that provides exact instructions about an event or a project. You can import either. If you use **File>Import>XML**, you get a dialog that allows you to select on which drive the imported project or event will be saved. Importing an XML file of an event will copy the media for that event into the new location. You should be aware that some folders and Smart Collections do not import when importing events.

Currently, this version of XML is in its infancy and needs to adopted by third-party providers who can interface with it. At the moment XML files exported from legacy versions of FCP cannot be imported using this XML function. Perhaps that will be available at a future date.

CONVERTING FILES

Although FCP can work with just about any type of file, not all files are equal; some are more equal than others. FCP likes QuickTime files, and while it can accept most formats and frameworks, some, such as MPEG-2, DV Stream, or AVI, should be converted before being imported.

There are a number of tools you can use to do this. If you have Compressor, you can use that to convert a great many file types to ones that are more convenient for FCP. Another excellent conversion utility is the free tool MPEG Streamclip, which can be downloaded from www.squared5.com. While Compressor cannot rip MPEG-2 files, MPEG Streamclip can also be used to convert many file types, including MPEG-2 and unprotected DVD content, into a format that can be edited in FCP. DVD conversion does require purchasing the MPEG-2 Playback Component from Apple. If you have an earlier version of Final Cut Pro, not Express, you already have this. Another useful tool is ffmpegX, which can be found at http://ffmpegX .com/index.html. A commercial package called Cinematize easily rips unprotected

DVDs in whole or, by selecting parts, into any QuickTime format desired (www.miraizon.com).

In MPEG Streamclip, if you need to convert a single file or combine multiple files into a single clip, use **File>Open Files**. If you need to convert multiple files in MPEG Streamclip, rather than using the Open Files function, use **List>Batch List** to access multiple files in a single folder.

Follow these steps to convert your MPEG-2 standard definition material:

1. Use **File>Export to QuickTime**, which brings up the window in Figure 3.20.
2. Set frame size on the left to 720×480 (DV-NTSC) or 720×576 (DV-PAL).
3. Set the frame rate to 29.97 for NTSC or 25 for PAL.
4. Set the **Field Dominance** to Lower.
5. Set the Compressor pop-up at the top to **Apple DVCPRO50-NTSC** or **Apple DVCPRO50-PAL**. These are excellent codecs, with twice the data rate of standard

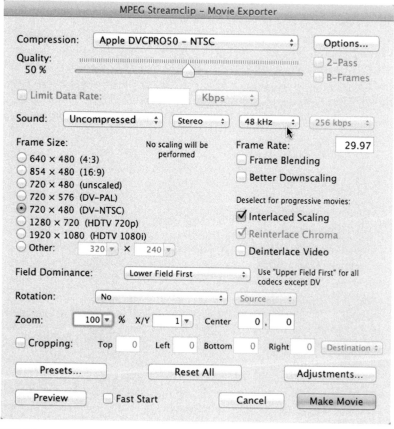

FIGURE 3.20

MPEG Streamclip Movie Exporter.

DV, but not as weighty as ProRes. You could also use Apple ProRes LT, which is
another excellent codec.

6. If your media is widescreen, click the **Options** button next to the compressor
type and change the **Aspect Ratio** to 16:9. If you wish, you can also make the
media progressive rather than interlaced by selecting it here.

7. Finally, set the audio to **Uncompressed Stereo 48kHz**.

Importing Music

The simplest way to import music is to use the music note button on the right side of
the Toolbar (see Figure 3.21). This will allow you access to your audio content that
comes with FCP and iLife, as well as your own iTunes library. To bring the music

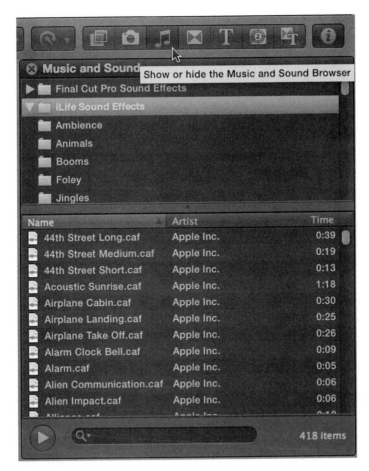

FIGURE 3.21

Music and Sound Browser.

into your project, simply drag it from the media browser into the Timeline. This makes a copy of the file inside the FCP project. You can also drag directly to an event.

This is all well and good in theory, but the fact is that a great deal of iTunes music is in heavily compressed formats like MP3 and, while they will work in the new FCP, they are not really suitable for professional production work and should be converted to high-quality AIFF files—even AIFF files such as those imported directly into iTunes without conversion are not in the best possible format. Audio CDs use an audio sample rate of 44.1kHz. This is not the sampling rate used by digital video, which is most commonly 48kHz. MP3s should also be converted to the AIFF format while being resampled and having their compression removed. There are a number of ways you can convert audio files. This can be done in the QuickTime 7 Pro Player. The standard QuickTime X player will not be sufficient, nor the standard QuickTime 7 player. By upgrading to the $29 Pro version, you get the ability to change files into several different formats.

You can also convert files using Compressor, which has an AIFF 48k preset that can be used or made into a droplet; a small applet that can be used to immediately convert files that are dragged onto it into whatever specification the applet uses.

If you do not have the QuickTime 7 Pro Player, you can also make the conversion in iTunes. To do this, you must set up your iTunes preferences:

1. Under the **iTunes** menu, go to **Preferences**.
2. In the first panel, **General**, next to the pop-up for **When you insert a CD**, click on the **Import Settings** button.
3. From the **Import Using** pop-up menu, select **AIFF Encoder**.
4. From the second pop-up menu, select **Custom**.
5. Set the **Sample Rate** to 48.000Hz.
6. Set the **Sample Size** to 16-bit.
7. Set the **Channels** to Stereo, as in Figure 3.22. Now you're ready to convert your music.

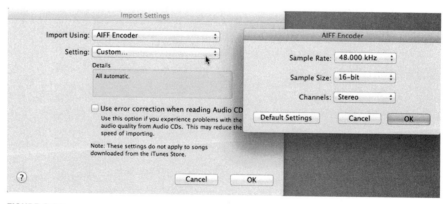

FIGURE 3.22

Channels Set to Stereo.

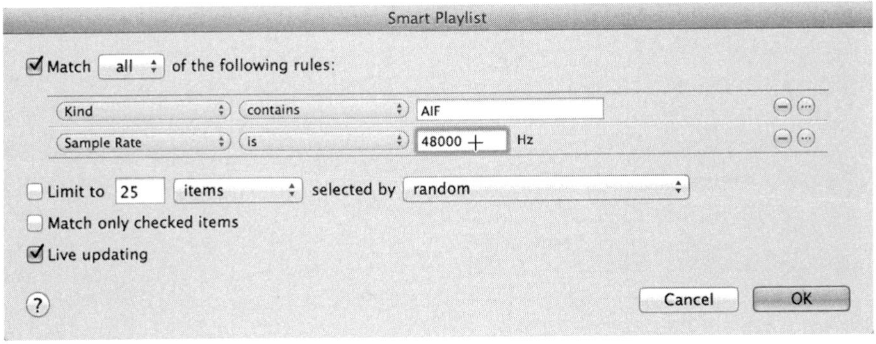

FIGURE 3.23

Smart Playlist Criteria.

8. In the **iTunes** window, select the tracks you want to convert by either **Shift-clicking** to select contiguous tracks in the list, or **Command-clicking** to select separate tracks.
9. With the tracks selected, from the menus go to **Advanced>Create AIFF version**.

iTunes will make an AIFF copy of the file in your iTunes library. iTunes can sometimes be a labyrinthine place to find a track. To make it easier to find your converted AIFFs, you can either make a playlist or a smart playlist.

1. Use **File>New Smart Playlist**.
2. In the criteria, set the first pop-up to **Kind**, the second to **Contains**, and in the box, type **AIF**.
3. Click the + button to add a second criteria. Make the first popup **Sample Rate**, the second popup **is**, and type in **48000** in the value box (see Figure 3.23).
4. Name the smart playlist whatever you want. The smart playlist will appear in your **iTunes** library list in FCP.

SUMMARY

With these last two lessons, you have completed the process of correctly setting up your system and the application, and we have gone through the importing process. First, we set up our preferences for editing, for playback, and for importing. We created a new event and looked at importing from file-based cameras and from tape-based cameras; and we created camera archives. We also imported still images from iPhoto and Aperture, libraries and projects from iMovie, and audio from iTunes. We also looked at converting files from DVD. We are finally ready to begin the editing process. To start, I suggest working with the tutorial media provided, or if you feel comfortable with the application, try working with your own media. Next, we'll look at organizing media in the Event Library using keywords and collections.

Organization

4

LESSON OUTLINE

In this lesson we'll look at working in the Event Library to organize footage using FCP's tools for working with metadata, the extra information attached to a piece of media that either comes from the camera itself or that you add to it.

LOADING THE LESSON

If you skipped directly to this lesson, first you must download the material from www .fcpxbook.com. For this lesson we'll need the ZIP file called *FCP2.sparseimage.zip*:

1. Download the file and then copy or move the entire ZIP file to the top level of your dedicated media drive.
2. Double-click the file to open the *FCP2.sparseimage*.
3. Double-click the sparse image to mount the disk.
4. Launch the application.

There is no projects folder on this disk, so your Project Library may be empty, but there is an events folder that you will see in the Event Library. It contains one event called *Organize* with 13 clips.

The media files that accompany this book are heavily compressed H.264 files. Playing back this media requires a fast computer. If you have difficulty playing back the media, you should transcode it to proxy media:

1. To convert all the media, select the event in the Event Library, **Control**-click on it, and choose **Transcode Media**.
2. In the dialog that appears, check on **Create proxy media**.
3. Next, you have to switch to proxy playback. Go to **Preferences** in the **Final Cut Pro** menu.
4. In the **Playback** tab, click on **Use proxy media**. Be warned: If you have not created the proxy media, all the files will appear as offline, so you will have to switch back to **Use original or optimized media**.

ORGANIZING THE CLIPS

There are no firm rules about how you organize your clips, and each project tends to dictate its own organizational structure. Usually, I begin with one event that holds all of the media for the production, video, audio, and stills, so they're all collected into one place and are easy to back up. I like having the complete event that allows me to skim through all the material toward the end of the project to make sure I haven't overlooked or discarded anything, which can be useful in light of the way the material gets cut together.

From the event, you start breaking down your material to organize it. Making copies of clips for organizing in the library does not duplicate the media on the hard drive and doesn't take up more space. From the master shots, which, in the *FCP2* disc, you can access by selecting the *Organize* event itself in the library, you can organize the clips into collections. These are essentially bins as were used in Final Cut Express and legacy versions of FCP, except that they are based on metadata, either that the application uses based on specified criteria, which are Smart Collections, or based on keywords applied by you, which are Keyword Collections.

Collections can be used in a variety of ways. Narrative projects tend to have material broken down into Scene Collections, with separate collections for different types of shots or characters, which are all grouped together in a folder, depending on how complex the scene is, which is often identified by a scene number on a head slate linked to a shooting script. Documentary projects tend to break the material down into subject matter: a collection for all of the forest shots, another for logging scenes, another for road work, another for weather, another for all of the interviews, another for sound, another for narration tracks, another for music, another for graphics. FCP can accommodate both structured story editing forms required in feature film editing and free association often required in scriptless documentary and anything in between.

The real trick is to break down your material into enough folders and collections so that your material is organized, but not so many collections that it becomes difficult to find material. As you copy clips into collections, add notes—lots of them. The more information you include on the clips, the easier it will be to find them.

It is critical that you cut up your shots and organize them into collections to work efficiently, particularly for long-form work, productions longer than 20 minutes or so, short projects with a lot of material, or productions that may go on for a longer period of time. The longer the production, the more cards you've archived, the more project timelines, the more complex everything becomes. Having your material well organized is crucial. One nice feature of the new FCP is its ability to handle very large projects. In previous versions, after the project file size reached a certain level, about 50MB or so, it was a good idea to split up the project. This is no longer the case with the new 64-bit version of Final Cut.

NOTE

Bins

Legacy Final Cut Pro and Final Cut Express had folder-like items that were called bins. Bins are gone in FCP. We now have collections, both Keyword Collections and Smart Collections, and folders. You cannot move clips into folders; you can only move them into collections by adding metadata to the clips, which can be done by dragging the clip into a Keyword Collection. The collections themselves can be organized into folders.

Naming Clips

Because of the extensive use of metadata in FCP, clip names are less important than they used to be in legacy Final Cut Pro and Express. Nonetheless, a cluster of date/time stamps like *2011-05-01 14:05:06* isn't very appealing or informative. Although you can rename your clips whatever you like, this does not change the name of the media file on your hard drive that it refers to.

You can rename a clip in the Event Browser in either Filmstrip view or in List view. To change the name in Filmstrip view, just click the name, and it will be selected for renaming. In List view, select the name and press **Return** to highlight the name and allow you to change it.

You can also change the name of a clip in the Info tab of the Inspector. Clips in the Timeline can also be renamed in the Inspector's Info tab, and unlike previous versions of Final Cut, this does not rename the clip in the browser. The two are quite separate, two individual instances of the same piece of media. However, within the browser, any instance of a clip in any collection or Smart Collection has to share the same name.

Name	Start	End	Duration	Content Created
2011-05-01 14:58:27	00:00:00;00	00:00:33;18	00:00:33;18	May 1, 2011 2:58
2011-05-01 14:59:05	00:00:00;00	00:00:16;10	00:00:16;10	May 1, 2011 2:59
2011-05-01 14:59:30	00:00:00;00	00:00:26;18	00:00:26;18	May 1, 2011 2:59
2011-05-03 14:05:52	00:00:00;00	00:00:12;04	00:00:12;04	May 3, 2011 2:05
2011-05-03 14:07:54	00:00:00;00	00:00:19;04	00:00:19;04	May 3, 2011 2:07
2011-05-03 14:28:33	00:00:00;00	00:00:10;22	00:00:10;22	May 3, 2011 2:28
2011-05-03 15:11:23	00:00:00;00	00:01:25;28	00:01:25;28	May 3, 2011 3:11
Bananas & coffee	00:00:00;00	00:00:17;22	00:00:17;22	May 1, 2011 2:21

FIGURE 4.1

Name Column in the Browser.

Let's start by renaming a few of the clips, first in Filmstrip view:

1. Change to Filmstrip view by clicking the icon at the bottom of the library or press **Option-Command-1**.
2. To see the clips as single icons, change the slider at the bottom of the browser to All or press **Shift-Z** for zoom to it, which sets the slider to the right.
3. Click the name of the first clip, the shot of a bucket of green coffee beans, to select it and rename it *Green beans*.
4. Let's switch to List view by clicking the List View icon or pressing **Option-Command-2**.
5. Select the next clip in the list, the shot of the coffee cherries on the tree, press **Return** to highlight the name and rename it *Cherries on tree*.
6. In whichever view you prefer, rename the next clip *Bananas & coffee*.
7. Select the clip *2011-05-01 14:39:59* and open the Inspector to the Info tab.
8. In Basic view, triple-click or drag-select the clip's name and change it to *Green beans on tray*.
9. The last clip we'll rename is *2011-05-01 14:56:18*. This is a two-minute shot at the end of coffee roasting process. Just call it *Roasting*.

If the Name column is selected at the top of the List view, as in Figure 4.1, the named clips will be at the bottom of the list. To reverse the sort order, just as in the Finder, click the Name header to toggle the triangle direction and reverse the order. If you want to keep the clips in shot order, regardless of the name, click the **Content Created** column header. This is sometimes useful for shoots like this that are a sequential process.

Deleting Clips

In legacy versions of Final Cut, you could simply select a clip and press the **Delete** key, and it would disappear from your project, but it would still be on your hard drive. In this version of FCP, selecting a clip and pressing **Delete** simply marks it as rejected, as we will see in a minute. Because the event you're working with actually contains your media, the clips in the Event Browser are very much your media. You can delete a whole clip from your event and remove it from your hard drive. Fortunately, you get a warning message when you do this, and if you click **OK**, it's gone. But if you have sudden second thoughts, **Command-Z** undoes this action, removing the clip from the Trash and restoring it to the browser (and to your Events folder in the Finder). Note that deleting from the sparse image is slightly different and is not undoable. You will get a second warning to this effect when deleting a clip from these disk images.

Select the clip *2011-05-01 14:55:28* and press **Command-Delete**. If the event contains the actual media, as it does here on the disc called *FCP2*, the clip is gone. If the event is holding only alias pointers to the media, the alias is deleted from the event, but the media remains wherever it was.

TIP

Restoring Original Clip Name

If you ever want to restore the original clip name after renaming it, do this:

1. Select the named clip, and **Control**-click to select **Reveal in Finder**.
2. Click once on the clip name in the Finder and copy it with **Command-C**.
3. In the Event Browser, highlight the clip name and use **Command-V** to paste in the Finder name.
4. Remove the **(id)** prefix and replace the underscores with colons if you wish. Colons are not allowed in the Finder but are fine in FCP.

NOTE

Trimming Rejects

There is currently nothing equivalent to trimming rejected portions of clips and moving them to the trash as there is in iMovie, nor does FCP have the trimming functions of Media Manager in legacy versions of Final Cut to consolidate your media.

KEYWORDS

Collections can be made by using keywords or by creating Smart Collections. This adds metadata to your clip. You can select an entire clip or a portion of a clip to be added to a collection. A shot can be long or short, but you'll usually want to use only part of it, perhaps cutting out the bit at the beginning where the camera's not steady or a section where someone steps in front of the lens. You can do this in a couple of ways in FCP: one familiar to iMovie users and one familiar to Final Cut users that should be used, as it allows for the precision needed in video editing.

FIGURE 4.2

Dragging a Selection.

1. Select the *Cherries on tree* clip. (It's Content Created time is May 1 at 2:15:30 p.m.)
2. Switch to List view if you are not in it and slide the pointer across the preview image. Make sure the preview is not selected and is not bounded by a yellow selection box.
3. Skim across the clip to look at the video. Around the 54-second mark, mouse down and drag a selection of about 6 seconds of the clip. As you do this, a tooltip will show you the length, as in Figure 4.2. All iMovie users know how to do this.
4. Once you have the selection made, you can drag the handles on either end to adjust it.
5. Let's drop the selection. You can do this by clicking off the preview or by pressing **Shift-Command-A** for Deselect All.
6. Play the clip using the **JKL** keys and find 54:04. Obviously, on a clip like this that's over a minute long, this gives you much greater precision.
7. Press the **I** key to mark an In point at 54:04. The **Left** and **Right** arrow keys will let you move forward and backward in one-frame increments. **Shift** and the **Left** and **Right** arrow keys will let you move forward and backward in 10-frame increments.
8. Play through the clip until 1:00:22 and use the **O** key to mark an Out point. This makes a selection, and the selected duration 6:17 appears at the bottom of the browser. This method of marking a selection will be familiar to Express and Pro users.

The best way to mark this section of the shot is to make it a Favorite or to apply a keyword to it.

> **TIP**
>
> Going to an Exact Frame
>
> You can use the Dashboard to go to an exact frame. Click the timecode in the Dashboard once, or press **Control-P** for playhead and type in a number such as *5404* for 54:04 (see Figure 4.3). You don't need to type in the colon. Press **Return** and the playhead will leap to that position.
>
>
>
> **FIGURE 4.3**
>
> Playhead Position.

> **TIP**
>
> Playing a Selection
>
> To play just a selection that you've made from beginning to end, press the forward slash (/) key. This is the equivalent of the Play In to Out function in Express and Pro, which was Shift-\. The new keyboard shortcut is easier, I think.

Anything that happens in an event in the Event Library is tracked by a database of information that's filed in the event folder and called *CurrentVersion.fcpevent*. This keeps track of all the metadata applied to all the clips. The whole organizational structure of FCP is based on the application of metadata to clips. Much of it is already there, from the camera and the QuickTime file or other file format that is your media; the rest you have to provide. One way you can do this is by adding keywords.

Keywords can be added to clips either in List view or in Filmstrip view, regardless of how the selection is made—either of part of a clip, as here, or of an entire clip. To add a keyword to a clip or section of a clip, follow these steps:

1. With the area selected, click the **Keyword** button in the button bar below the Event Browser or by pressing the keyboard shortcut **Command-K**. Either method will open the keyword HUD (Heads Up Display). **Command-K** will also close it.
2. Enter the words *cherries* in the HUD and press **Return**. An animation will appear that shows the keyword is applied to that segment of clip, and a blue bar will appear on it. Also in the *Organize* event, a new Keyword Collection called *cherries* is created (see Figure 4.4), and the keyword is added to the clip, which can be seen in List view in Figure 4.5.

FIGURE 4.4

New Keyword Collection.

FIGURE 4.5

Keyword in List View.

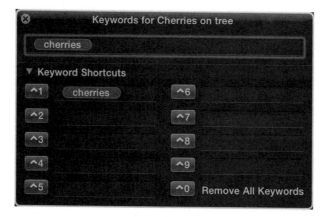

FIGURE 4.6

Keyword HUD.

3. Twirl open the disclosure triangle to see the rest of the HUD, as in Figure 4.6. Notice that the keyword has a keyboard shortcut, **Control-1**, applied to it.

4. Click on the *cherries* Keyword Collection in the Event Library, and you will see that it contains just one clip: the section of the *Cherries on tree* clip.

TIP

Opening Disclosure Triangles

In the Event Library, Event Browser, or Project Library, you can select an item or items and use the **Right** arrow to twirl open the disclosure arrow for the item. The **Left** arrow will close the disclosure triangle for selected items.

You can have up to nine keyboard shortcuts for keywords at a time using **Control-1** through **9**. The last shortcut, **Control-0** (zero), is reserved to allow you to select a clip and remove all the applied keywords. You can also type a keyword in HUD to add it to the list. This capability is useful if there are commonly used keywords that you want to apply. You can enter them even before you import your media. After you've made the keywords, as soon as you want to apply one, you just press the keyboard shortcut. If it's the first time the keyword is applied to a segment or a clip, a new Keyword Collection will be created in the Event Library.

If you like clicking buttons, you can also click the shortcut button in HUD to apply a keyword to a selection.

Once you've made the Keyword Collection by adding a keyword to a clip, it becomes easy to apply the same keyword to other clips or to other segments of the same clip:

1. Click back on the *Organize* event in the Event Library so you see the whole list of clips.
2. Start playing the *Cherries on tree* clip from the beginning and mark an In point around 17:07 after the camera has moved in to the green cherries.
3. Play forward through the push in and mark an Out point at 23:08; then press **Control-1** to apply the *cherries* keyword.
4. Let's add another keyword. Play forward to 43:21 and mark an In point. Mark an Out point after the push at 50:27.
5. Put the pointer over the selection till you see the helping hand. Grab the selection and drop it on the *cherries* Keyword Collection in the Event Library. The keyword is applied to that section.
6. Click off the clip preview in the browser to drop the selection and open the disclosure triangle next to the clip name. Then click the first *cherries* keyword in the List view. Notice the selection area you made and labeled shows in the preview.
7. Click the second blue keyword bar in the preview, and that section is immediately selected.

Essentially, marking keywords as shown here is making multiple In and Out points in a single clip and holding them for any time you need them. This is analogous to, although not quite the same as, the Express and Pro method of making subclips. There are no master clips as such in FCP, as there were in the older versions, so these are not subclips in the sense that they cannot be renamed separately. But in the *cherries* collection, each keyworded section behaves as if it were a separate clip, and like subclips, they are not limited to that section of media. Keyword selections can be expanded to the original clip length simply by dragging them out in the Timeline, whereas in legacy Final Cut, the subclip limits had to be removed by a special command, which reset the subclip. Although making a keyworded clip or a Favorite, which we'll see in a moment, does not create a separate QuickTime file, the clip in the Keyword Collection is treated as a separate clip, even though it really isn't.

If you worked in Final Cut Express or legacy Final Cut Pro, you may remember creating and working with subclips. One of the essential capabilities of subclips was that they were master clips and could be renamed. Clip names were an important part of finding material in the application. In FCP, the clip name is perhaps the least important piece of metadata about it. These names have been superseded by keywords and other collections.

Let's see how to remove a keyword:

1. In the preview in List view of *Cherries on tree*, skim down to about 1:06:15 and drag a selection that's about eight seconds long.
2. Press **Control-1** to add the *cherries* keyword. Oops. Those aren't coffee cherries; they're papayas.
3. If you dropped the selection, click the last blue bar on the clip or click the last *cherries* keyword in the list.
4. If the keyword HUD isn't open, press **Command-K**, which will also close it.
5. Select the word *cherries* in the HUD, as in Figure 4.7, and press the **Delete** key. The keyword is removed and disappears in a little puff of smoke. The section is, of course, also removed from the Keyword Collection.

You can also create a Keyword Collection and import directly into it:

1. To create a new Keyword Collection, you can use **File>New Keyword Collection** or the shortcut **Shift-Command-K**. Or you can **Control**-click on the event in the library and select **New Keyword Collection**.
2. Choose your preferred method to make a new Keyword Collection and call it *roasting*.
3. In the Event Browser, you have two buttons that you can use to import clips directly into that Keyword Collection (see Figure 4.8).

FIGURE 4.7

Keyword Selected for Deletion.

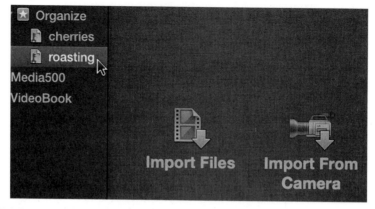

FIGURE 4.8

Empty Keyword Collection.

If you wish, you can also drag and drop clips, video, audio, or stills directly from the Finder into a Keyword Collection. The items become keyworded and are added to the event the collection is in.

NOTE

Across Multiple Clips

You cannot drag-select or mark In and Out points across adjacent clips in either List view or even in Filmstrip view.

FAVORITES AND REJECTS

Next to the **Keyword** button in the Toolbar is another group of buttons for Favorites (the green star), Unrated (the empty star), and Rejects (the red X). Let's mark some Favorite sections on the clip we named *Roasting*, which is over two minutes long. Stretch out the Event Browser so you can make it as wide as you can while still leaving the Event Library open. Then follow these steps:

1. Play the *Roasting* clip from the beginning and mark an In point at 4:10; then mark an Out point at 18:08 after the words "a medium roast."
2. To mark it as a Favorite, press the **F** key. A green bar appears in the preview, and the section is tagged a Favorite, as in Figure 4.9.

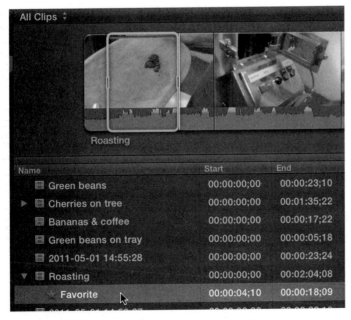

FIGURE 4.9

Favorite in Preview and List View.

FIGURE 4.10

Event Browser Filter.

3. Mark another In point at 22:06 and an Out at 34:26 after the words "from here to here."

4. This time, to mark the Favorite, click the green star in the Toolbar.

Notice that no collection is being created for Favorites, but we can fix that so you can have a collection for Favorites. We'll make a Smart Collection for that in a moment. To see just your Favorites, use the Event Browser Filter pop-up menu at the top left of the Event Browser (see Figure 4.10) or press **Control-F**. Only the two Favorite sections will appear. Press **Control-C** to return to all the clips in the browser. With the browser in List view, let's rename the Favorites:

1. Select the first *Favorite* in the list and press the **Return** key to rename it.

2. Type in *medium roast* and press the **Return** key to close the name.

3. Press the **Down** arrow to select the second *Favorite*; then press **Return** to rename it.

4. Type in the word *darker* and press **Return** again to confirm it.

Because you can't name the sections of the clip separately, naming the Favorites is a way to make a note of the section that you have marked as a Favorite.

Let's mark some sections of the clip as rejected:

1. Select the *Roasting* clip so that you can see the preview; then click the first of the two green bars to select the first Favorite segment.

2. Press **Shift-I** to move the playhead in the preview to the beginning of the marked selection and press **O** to mark an Out point. This will make a selection from the beginning of the clip to the beginning of the first Favorite.

FIGURE 4.11

Reject in Preview and List.

3. Press the **Delete** key to mark the section as rejected. A red bar will appear at the top of the clip showing that it's rejected. The rejected section will also display in the list under the clip, as in Figure 4.11.

4. Click the second green Favorite bar to select the region and press **Shift-O** to move the playhead to the Out point.

5. Press **I** to mark an In point and then play to around 42:00 and mark an Out point.

6. This time click the red X button to mark it as rejected.

7. Go up to the Event Browser Filter pop-up in the upper left again, and select **Hide Rejects** or press the shortcut **Control-H** for hide.

With rejects hidden, *Roasting* appears as two separate clips: one with two Favorites and one without any Favorites. These are not two separate clips that can be renamed. These Favorites are still one clip shown as two separate sections.

1. To return to viewing all the clips, press **Control-C** or use the pop-up.
2. To remove a Favorite or a Reject, click the green or red bar to select the region.
3. Press the **U** key for unrated or click the empty star in the Toolbar. The green or red designation disappears.
4. Undo that so that you still have two Favorite sections and two Reject sections in *Roasting*.

We've marked Favorites and Rejects and removed them. We've seen how the filter pop-up lets us change what's displayed in the browser. In addition to show All Clips and Favorites and Hiding Rejects, the filter can also be used to display only Rejects or to display segments of clips that have neither any ratings nor any keywords. This doesn't mean that every frame of every clip has to be keyworded or favored or rejected, just that these are segments you want to save or you would rather hide.

Search Filter

In addition to keywords and ratings, it's often useful to include notes about clips or segments on the clips themselves, a few words about a soundbite, or a note about fixes such as color correction that need to be made. Here's how:

1. If you're in Filmstrip view, switch back to List view by pressing **Option-Command-2**.
2. Make sure the *Organize* event is selected, click on the clip named *Green beans on tray*, and press the **Tab** key, which is another way to select the name for renaming.
3. Continue pressing the **Tab key** until the selection moves across the list to the Note column.
4. Type in the word *sample*. This is now a searchable piece of information that can be found easily and used as a tag for creating a collection.

To search the browser, click in the search box in the upper right of the Event Browser to enter a search word like *sample* (see Figure 4.12). This is a search filter, so you will now see only the clip *Green beans on tray*. If you select the Keyword Collection *cherries*, the browser will be empty because there is nothing in that collection that includes the word *sample*. To remove the text filter, delete the word or click the little X in a circle on the right edge of the search box.

You can call up a more robust filter window by clicking on the Magnifying Glass icon on the left edge of the search box, or just like in the Finder, pressing **Command-F** to bring up the filter HUD. Here, you can enter a text search such as *sample* and add multiple criteria by clicking the + pop-up in the upper right (see Figure 4.13). Notice the button allowing you to create a Smart Collection.

FIGURE 4.12

Text Search.

FIGURE 4.13

Search Filter HUD.

SMART COLLECTIONS

Metatagging is tagging media with metadata, such as we do when we create keywords or add notes to clips. This metatagging is used to create what Apple calls *rules* for Smart Collections. A Smart Collection is like a Smart Album in iPhoto or a Smart Playlist in iTunes. Basically, you set up criteria that the collection uses to add material to it. There are two types of Smart Collections: one you produce, which we'll see in a moment, and one produced by the application, either on importing or when analysis is run. FCP has some automated tools that divide material into shot types and shots with people. While this sounds good in theory, for large projects it's not really of great benefit, and you might be better off imposing an organizational structure that suits your project and your needs. Some Smart Collections are created automatically on import. These are grouped in the *Smart Collections* folder, in which you can find other folders for the *Shot Date*, *Shot Type*, and *People*. Inside the *Shot Date* folder are Smart Collections based on the date the media was shot. In *Shot Type*, there are Smart Collections based on the type of shot—medium shots, wide shots, close-ups. In *People*, there are Smart Collections for shots with one person, two persons, or groups. These are created on import, with the application analyzing the content and putting material in the appropriate Smart Collection. The same shot can appear in multiple Smart Collections: It might be in close-ups, one person, and, of course, in the shot date Smart Collection.

You can have FCP analyze your video either when setting up your import actions or after the clip is imported by using **File>Analyze and Fix** or by **Control**-clicking on a clip or selected clips and choosing **Analyze and Fix** from the shortcut menu. Here's how:

1. Select the clip *2011-05-03 15:11:23* and rename it *removing parchment from beans*, because the green coffee beans are pounded to remove the outer parchment-like skin that surrounds them. (This book is multi-educational.)
2. With the clip selected, **Control**-click and run **Analyze and Fix**. In the dropdown menu that appears, check on Analyze for stabilization and rolling shutter, Analyze for balance color, Find People, and finally check on Create Smart Collections after analysis, as in Figure 4.14.

This process may take quite some time, depending on the speed of your computer and what else you're doing on it. To see the progress of your analysis, watch the dial on the left of the Dashboard or click it to bring up the Background Tasks

FIGURE 4.14

Running Analysis and Creating Smart Collections.

HUD (see Figure 4.15). You can also bring up this HUD with the keyboard shortcut **Command-9**, which also will close it.

Once the analysis is complete, you will have a number of new collections with the purple gear icon called Smart Collections: three grouped in a folder for *People* and two in a folder for *Stabilization* (see Figure 4.16). On your hard drive in the event folder, you will see a new folder called *Analysis Files* with other folders inside it.

FIGURE 4.15

Background Tasks HUD.

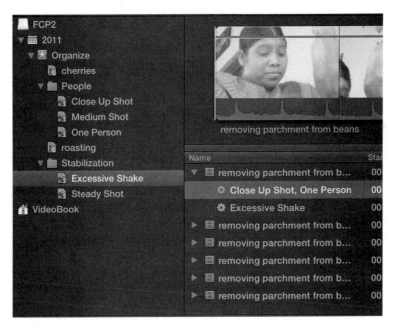

FIGURE 4.16

Smart Collections.

If you select the *Excessive Shake* collection, it will show you the six pieces that have been clipped out of the shot that meets its criteria.

NOTE

Renaming Collections

While Keyword Collections can be renamed, thus changing the keyword itself, automatic Smart Collections *cannot* be renamed, as they take their name from the analysis that was performed. Renamed Keyword Collections will change every instance of the keyword used in the browser. Renaming will not change keywords that have already been edited into a project. On the other hand, *manually* created Smart Collections can be named whatever you want, and renaming them will not change the criteria.

Obviously, you can run multiple clips or all your clips through these analyses, and you can continue working while this automatic process continues. There are a few functions you cannot do while this is going on, however. You cannot rename the clips. You cannot delete the clips from the event. You cannot quit the application. You can pause or stop the processes, and these can be picked up at a later time, but generally it's not a good idea to interrupt the analysis.

These automatic Smart Collections are created by FCP. You can delete a Smart Collection at any time. Simply select it and press **Command-Delete**. If you do this, the Smart Collection is gone, but the analysis and the analysis keyword remain attached to the clip, as in Figure 4.17.

You can remove any one of the analysis keywords by **Control**-clicking on it and selecting **Remove Analysis Keywords**. This will remove just the selected keyword. If you want to remove them all, select them all and use the shortcut menu. Alternatively, simply select the clip itself and press **Control-Shift-0** (zero), and they will all be gone. Neither analysis collection keywords nor the Smart Collections can be renamed.

▼ 🖼 removing parchment from beans	00:00:00;00	00:01:25;28	00:01:25;28
✿ **Excessive Shake**	00:00:03;05	00:00:03;27	00:00:00;22
✿ Excessive Shake	00:00:17;09	00:00:28;12	00:00:11;03
✿ Medium Shot, One Person	00:00:28;00	00:00:30;00	00:00:02;00
✿ Excessive Shake	00:00:28;25	00:00:29;22	00:00:00;27
✿ Excessive Shake	00:00:30;27	00:00:31;20	00:00:00;23
✿ Close Up Shot, One Person	00:00:46;00	00:00:52;00	00:00:06;00
✿ Excessive Shake	00:00:50;14	00:01:03;07	00:00:12;21
✿ Excessive Shake	00:01:20;06	00:01:25;28	00:00:05;22

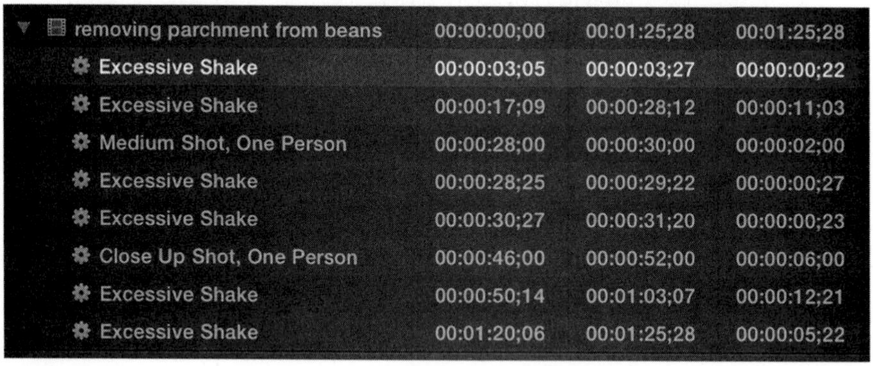

FIGURE 4.17

Analysis Keywords.

You could make an empty Smart Collection by using **File>New Smart Collection** or the shortcut **Option-Command-N**, and then double-click the icon to initiate the Search Filter HUD. But let's do it another way:

1. Select the *Organize* event in the Event Library so you'll be searching all the clips, and then click the Magnifying Glass icon on the right side of the search box to bring up the Search Filter HUD.

2. Type the word *sample* in the Text search box. If you entered the word *sample* in the Note column earlier, the filter will reveal the one clip in the browser called *Green beans on tray*.

3. Next click the + button in the upper right of the Search Filter HUD and select Ratings. The default is Favorite, and, of course, the browser will be empty because there are no clips that are both Favorites and have the word *sample* in them.

4. Change the pop-up in the upper left from All to Any, as in Figure 4.18. Now you'll have three clips or three segments of clips.

5. In the lower right, click the button for **New Smart Collection**. The criteria you used to make the smart search filter is used to make a new Smart Collection that appears in the event as *Untitled*.

6. Name the Smart Collection *Favorites+*. Now every time you make a Favorite or add the note *sample* to a clip, it will immediately be added to the Smart Collection.

Search filters are based on Boolean selections, so, for instance, you can search for text that Includes or Does Not Include or Is or Is Not. You can search a date based on when the clip was created or when it was imported, on a set date, before a date, after a date, or whether it was the last *x* number of days, or not the last *x* number of days.

FIGURE 4.18

Search Filter HUD.

FIGURE 4.19

Keyword Search Filter.

As you create the filter, the browser will update to your specified selections. Any time you create clips that meet these criteria, the clips, or selections, will automatically be added to the Smart Collection.

You can also use the Search Filter to create Smart Collections of multiple keywords. Create a Smart Collection, and from the + pop-up menu, select **Keywords**. All the keywords you've applied will appear. With the checkbox, you can choose **Uncheck All** (see Figure 4.19) and then check on the ones you want.

> **NOTE**
>
> Spanned Searches
>
> You can search multiple events at one time if you want by selecting all of them, but you can make a Smart Collection for only a single event. The Smart Collection has to be in one event and cannot bridge multiple events.

OTHER TOOLS

Just as the analyses of Smart Collections such as *Close Up Shots, Medium Shot*, and *One Person* are grouped into a folder, you can group your own collections into folders by **Control**-clicking on the event and selecting **New Folder**, or with the event selected, use **File>New Folder** or press **Shift-Command-N**.

You can also consolidate events by dragging one event on top of another in the Event Library. This will move all the media into one event folder. You have to do this inside FCP so the application knows that the files used in projects have been moved. You can also copy an event from one drive to another, and you can move an event by using **File>Move Event** or, just as in the Finder, by dragging to another drive while holding the **Command** key.

Compound Clips and Auditions

Compound Clips and Auditions are some advanced tools, which we'll look at later in situations in which the developers intended they be used, but FCP users have found them useful as organizational tools, so I've included them here as well. In the browser, even in Filmstrip view, it's difficult to play through your media because it stops at every clip. You can use the apostrophe (') key to jump to the next clip in Filmstrip view and start playback again, or you can use the **Down** arrow key to go down through the clips in List view to start playing that clip. But wouldn't it be great to group a bunch of clips together into the Timeline and just sit back and look through them? This is what Compound Clips can do. Here's how:

1. Select the *Organize* event and make sure nothing is selected in the Event Browser. Then use **File>New Compound Clip** or **Option-G**. You can also **Control**-click on a clip in the Event Browser and select **New Compound Clip**. When you do, you get a dialog asking you to name the Compound Clip.
2. Name it *Coffee*. If nothing was selected, an empty Compound Clip icon will appear in the preview, as in Figure 4.20.
3. Double-click this icon to open it into the Timeline, but then undo that with **Command-Z**. Rather than making an empty Compound Clip, let's combine all our clips into a single Compound Clip, or all the clips from a collection.
4. In the *Organize* Event Browser, press **Command-A** to select all the clips.
5. Then press **Option-G** to make a new Compound Clip.
6. Name the new Compound Clip *Coffee*. The new Compound Clip will appear at the top of the list, as in Figure 4.21. Notice the little badge icon in the upper left of the Compound Clip's preview.
7. Double-click the Compound Clip to open it into the Timeline. Here are all the clips laid out in the order in which they were shot for you to see. Press **Shift-Command-F** for full screen and sit back and watch your video play.

FIGURE 4.20

Compound Clip Icon.

FIGURE 4.21

Compound Clip in Browser.

This is actually a separate sequence Timeline inside your event. You can edit it as you'll see in the next lesson. You can add music and titles and effects. You can cut and paste from it into a project. You can place the entire Compound Clip into a project if you wish. If you're coming from Final Cut Express and legacy Pro and are concerned about not having multiple sequences inside a project, here are multiple sequences inside a single event. This is one way of keeping multiple sequences, each event a separate scene, each scene with its own compound clip.

It's important to note that although Compound Clips may behave as separate sequences, they are not fully recursive. That means if you make a Compound Clip in the Event Browser and put that Compound Clip into a project, the two are separate. If you make changes to the Compound Clip in the Event Browser, the Compound Clip in the project will not change. Similarly, any changes made to the Compound Clip in the project will affect only that project and will have no effect on the Compound Clip in the Event Browser.

Auditions are slightly different. They allow you to combine multiple clips into a container from which you can select the clip you want to view and use. When you edit the Audition into the Timeline, all the clips go together, and in the Timeline, you can switch between them. We'll look at working with Auditions in the Timeline in later lessons, but let's look at creating them here:

1. In the *Organize* event, select the three clips *Cherries on tree, Bananas & coffee,* and *Green beans on tray.*
2. To make the Audition, you can either **Control**-click on the clips or select **Create Audition** from the shortcut menu. A new item will appear in the browser

with a Spotlight icon, which can also be seen as a badge in the upper-left corner of the preview, just like the Compound Clip badge (see Figure 4.22).

3. Click the badge in the upper-left corner of the preview to open the Audition container (see Figure 4.23).

You can switch between the Audition clips in the container by clicking on the clip you want, and the preview will update. You can also move between the items using

FIGURE 4.22

Audition in the Browser.

FIGURE 4.23

Audition Container.

the **Left** and **Right** arrows. To close the Audition, you can click **Done** or press the **Y** key. The **Y** key will also open the Audition if you have it selected in the Browser. The Audition is a complete unit, separate from all the clips, and contains the clips you want in it. This is a handy way to group clips together, which can be edited and switched as you like. It's a great tool for prevaricators or those suffering from a serious case of indecision.

You can edit the Audition into the Timeline, where you can continue to switch between items. Once the Audition is in the Timeline, more clips can be added to it, or clips can be removed from it.

Though we've made the Compound Clips from complete clips here, if you have a group of selections in a collection, you can combine these edited clips into either Compound Clips or Auditions.

SUMMARY

In this lesson, we looked at the importance of organizing our material. We looked at naming our clips and deleting clips. We created keywords, made Keyword Collections, made Favorites and Rejects, and organized our event in collections and folders. We used the search filter to make Smart Collections. We also looked at Smart Collections created by analysis. We even worked with some advanced editing tools, Compound Clips and Auditions, and used them for grouping and organizing our media. In the next lesson, we will look at putting our shots together in a sequence. We'll look at precise ways of editing our clips into a project, as well some basic trimming and rearranging functions.

Editing Basics: The Edit Functions

5

LESSON OUTLINE

Now that you've organized your material in the Final Cut Pro Event Browser, it's time to edit your clips into the Timeline to assemble your project. FCP has several different ways to perform most of the editing functions. You can edit directly into your project Timeline with the mouse or with keyboard shortcuts, which is probably the most efficient way to edit in most instances.

LOADING THE LESSON

If you skipped directly to this lesson, first you must download the tutorial material from the www.fcpxbook.com website. For this lesson, we'll need the ZIP file called *FCP3.sparseimage.zip*. Proceed like this:

1. Download the file and then double-click it to open the *FCP3.sparseimage*.
2. Copy or move the entire sparse image file to your dedicated media drive.
3. Double-click the sparse image to mount the disk.
4. Launch Final Cut Pro.

On the sparse disk, you will find a couple of projects in the Project Library, which are actually for the next lesson, together with an event called *Edit* in the Event Library. Inside the event, you'll see the media we're going to work with in this lesson.

Remember, if one of the sparse images you're working on ever gets corrupted, or you need a new copy of something, you should either download a fresh copy from the www.fcpxbook.com website or unzip the downloaded file, move it to your hard drive, replacing the corrupted file, and relaunch it to get going again.

The media files that accompany this book are heavily compressed H.264 files. Playing back this media requires a fast computer. If you have difficulty playing back the media, you should transcode it to proxy media:

1. To convert all the media, select the event in the Event Library, **Control**-click on it, and choose **Transcode Media**.
2. In the dialog that appears, check on **Create proxy media**.
3. Next, you have to switch to proxy playback. Go to **Preferences** in the **Final Cut Pro** menu.
4. In the **Playback** tab, click on **Use proxy media**. Be warned: If you have not created the proxy media, all the files will appear as offline, so you will have to switch back to **Use original or optimized media**.

LOOK BEFORE YOU CUT

Whether you work your video into clips with keywords or with Favorites, you're really looking through your material. You should watch for relationships—shots that can easily be cut together. Getting familiar with the material is an important part of the editing process, learning what you have to work with and looking for cutting points.

Look through the shots in the Event Browser. Start at the beginning and play the clips rather than skimming them. Skimming is great if you need to work quickly and while you're organizing, but it's not really a good way to view your video, which should be done while viewing in real time. The Event Browser shows a few shots that have obvious relationships.

With the Event Browser set to sort by Content Created date and time, take a look at the last three video shots: *man standing at counter, man handed bag,* and *ls woman getting sample.* The same man is in the first two shots. A woman crosses through the second shot, and then she appears in the last shot. These shots can obviously be cut together to make a little sequence. Sometimes you might need a cutaway to bridge an edit where the shot change would occur, such as if you cut out a section in the middle of one of the shots.

Searching for these relationships between shots is critical as you look through your material. Some editors like to immediately create small sequences and group them together, not finely honed but roughly laid out so that first important impression is preserved. We saw one way to do this in the preceding lesson using Compound Clips. You may not use this approach in your final project, but assembling related shots quickly into a sequence is an efficient way to make notes about your material.

> **NOTE**
>
> The Cutaway
>
> Any editor will tell you that cutaways are the most useful shots. You can never have too many, and you never seem to have enough. No self-respecting editor will ever complain that you have shot too many cutaways. A cutaway shot shows a subsidiary action or reaction that you can use to bridge an edit, like the shot of the interviewer nodding in response to an answer. The cutaway allows you to bridge a portion of the interviewee's answer where the person has stumbled over the words or has digressed into something pointless. Often you can use a wide shot that shows the whole scene as a cutaway. Make note of these useful shots as you're watching your material.

EDITING CLIPS

Unlike in legacy versions of Express and Pro, in FCP you can have media ready to edit and no project or sequence to edit it into. The old software always started out with one empty sequence—not so here, where you make a project only when you're ready to start editing into a project. Let's start by making a new project. You can make a new project on any drive you have mounted and can see in the FCP Project Library. Normally, you would put your projects on your media drives, but for these lessons you can put the new projects on your system drive, which will put them into your home *Movies* folder. This might help you keep them separate from your own material.

1. To make the new project, select the drive in the Project Library. If the Project Library is not open, press **Command-0 (zero)**.
2. Select your system drive and make the new project by using **File>New Project** or the almost universal keyboard shortcut for a new document or whatever your application calls it: **Command-N**.
3. Name the project *Edit*, also setting the default event to *Edit*.
4. Leave the Starting Timecode at zeros.
5. Usually, the best choice, unless you have specific reason to use custom settings, is to have the video properties **Set automatically based on first video clip**.
6. Unless you're working in 5.1 surround and monitoring 5.1 surround while you edit, you should use a custom setting for the audio, as in Figure 5.1 with audio set to Stereo.

> **NOTE**
>
> Starting Timecode
>
> It's tradition for editors to use one hour (01:00:00:00) as the starting timecode for projects, or even earlier to include added bars and tone, program slate, and countdown leading to one hour exactly for program start. This timecode allowed editors to lay off back to tape without problems. There is no longer tape output in FCP. This may not be a problem for you. But if you may be going to tape at some future time, using third-party software, setting the start time to one hour might be useful. The QuickTime file exported from FCP with one hour as the start time will start at 01:00:00:00.

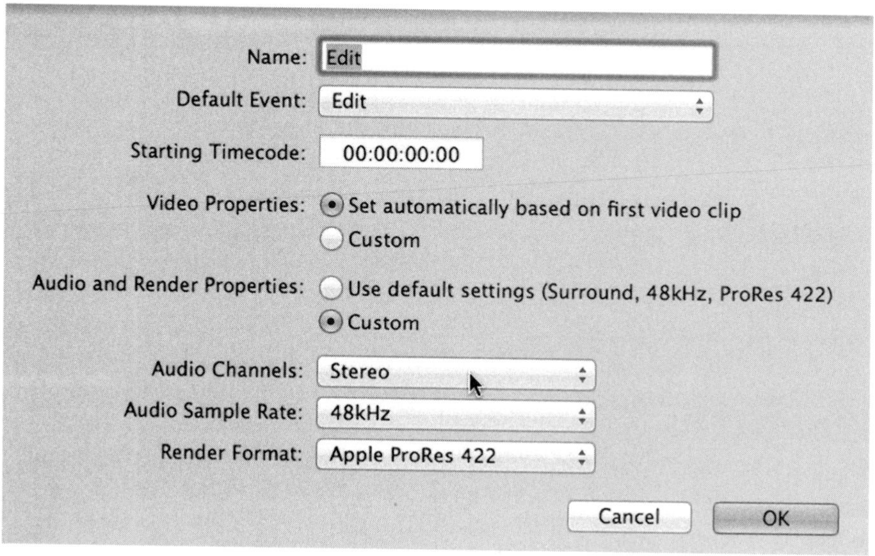

Name:	Edit
Default Event:	Edit
Starting Timecode:	00:00:00:00
Video Properties:	● Set automatically based on first video clip
	○ Custom
Audio and Render Properties:	○ Use default settings (Surround, 48kHz, ProRes 422)
	● Custom
Audio Channels:	Stereo
Audio Sample Rate:	48kHz
Render Format:	Apple ProRes 422

Cancel   OK

FIGURE 5.1

New Project Settings.

The new project will open in the Timeline panel named after your project—in this case, *Edit*. Notice the dark band along the middle of the window. This is your *primary storyline*. We'll start by editing some shots into this project. We've already seen some shots that work together, so we'll start there. For most types of edits, there are up to four ways to edit into the Timeline:

- By dragging the clip into the Timeline
- By using the **Edit** menu
- By using keyboard shortcuts
- By clicking one of the edit buttons

You can execute the edit any way you like. Many people like to drag to the Timeline, but I personally prefer the accuracy and exactness of using keyboard shortcuts.

Append

Let's start with *man standing at counter.* You could skim the shot to view it, which is fine when you're organizing your media and applying keywords; but when you're editing, to get the correct pace and timing for a shot, it's best to simply play it in real time. Here's how:

1. Play the shot with the **spacebar**. Wait for the camera to settle, and around 3:05, enter an In point with the **I** key to mark the beginning of the selection.

2. Play forward in the clip for around six seconds to around 9:10.
3. Enter an Out point by pressing the **O** key to mark the end of the selection.

TIP

Marking Favorites

A trick that many editors have adopted is to mark a selection with the **F** key, marking it as a Favorite, before editing the clip into the project. This technique makes it easy to come back to a selection you've already edited into the project, and marking Favorites allows you to make multiple selections to clips and save the selection information with the clips in the Event Browser. You can always select the entire clip as a Favorite as well and make fine-tuning decisions in the timeline.

Try it a few times until you're comfortable with the pacing of the movement. You might find that the more times you try it, the more you're shaving off the shot. Perhaps you'll feel that the front needs to be shortened as well. Don't cut it too tightly. If you find you're making it shorter and shorter, which is a common tendency, go back and make it longer. This is a rough selection, not finely tuned; that will be done once the clip is in the Timeline, where you can see the shot in context of neighboring shots. When you have it the way you want, about six seconds long, you're ready to put it into the Timeline.

The first way is to drag it there. Grab the image from the preview window in the Event Browser with the helping hand, as in Figure 5.2, and drag it into the Timeline. Drop it anywhere in the Timeline panel, and the clip will move to the beginning of the project.

man standing at counter

FIGURE 5.2

Dragging Clip to the Timeline to Append.

This type of edit always adds the clip to the end of the project. It doesn't matter where the playhead is, or the skimmer. If you simply drop the clip into the end of the Timeline, beyond any clips that are there, or into an empty Timeline, without placing it in the primary storyline, the clip will go to the end of the project. An edit that puts the clip at the end of the project is called an *Append* edit.

TIP

Marking Edit Points

Many editors like to mark their Out points on the fly. This allows you to judge the pace of the shot and the sequence—doing it in an almost tactile way—to feel the rhythm of the shot. After a few tries, you'll probably find that you're hitting the Out point consistently on the same frame. Marking the In point is a little different because often you want to mark the edit point before an action begins, but judging how far in front of the action to begin on the fly is difficult. Many editors like to mark the In point while the video is playing backward. By playing it backward, you see where the action begins, and you get to judge the pace of how far before the action you want the edit to occur.

NOTE

The FCP X Timeline

The FCP Timeline is different from earlier versions of Final Cut Pro and Express in that the Timeline is sort of a free-form palette for assemblies. It does not have distinct, marked tracks for video and audio, or a source-target patch panel—but rather an adaptable space that adjusts as needed for the media that's added to the window. The video and audio of synced clips appears together in the Timeline space, as in Figure 5.3.

FIGURE 5.3

Video and Audio in the Timeline.

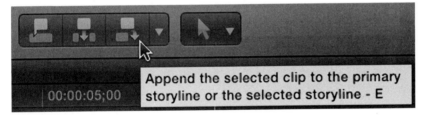

FIGURE 5.4

Append Button.

To get a better idea of what an Append edit is, let's edit a few more shots into the Timeline using other methods:

1. Let's put in the next shot, *man handed bag*. Play the clip in the Event Browser until the camera pushes in and settles around 2:19; then mark the beginning of the selection with the **I** key.
2. Mark the end of the selection with the **O** key at around 16:00 after he has his grocery bag.
3. To append the clip into the Timeline, use **File>Append to Storyline**. We have only a primary storyline at the moment, so the clip gets added to the end of project on the primary storyline, the dark band in the middle of the project.
4. Let's add the third shot, *ls woman getting sample*. Because the motion has already started at the beginning of the shot, play forward till about 9:25, after she gestures to the woman beside her. Mark an Out point for the end of the selection with the **O** key. The clip is selected from the first frame to the marked Out point.
5. To edit the clip into the Timeline, click the **Append** button, the third in the group of three in the Toolbar (see Figure 5.4).
6. Let's put in another piece of the same shot. Play forward to mark a selection starting at about 14:00, after the camera comes around to be beside her.
7. Play forward to about 24:07, shortly before the end of the shot; then mark the end of the selection.
8. Press the **E** key to do an Append edit and drop the clips in the Timeline after the others.

Insert

While the Append function is fairly common at the start of the editing process, very often you don't want to put the clip at the end of the project but somewhere in the middle of it. To do this, you do an *Insert* edit. Let's put a few clips at the beginning of the Timeline. Look at the lengthy shot called *cutting meat*. We're going to use a few pieces of this shot.

> **NOTE**
>
> The Magic Frame
>
> When the playhead moves to the end of the last clip in a project, the Viewer displays the last frame of the clip with a torn page overlay on the right side (see Figure 5.5). This is the Magic Frame because the playhead is actually sitting on the next frame of video—the blank, empty frame—but the display shows the previous frame.
>
>
>
> **FIGURE 5.5**
>
> Last Frame Indicator.

1. I'd like you to start with the wide shot, but because the shot is so long, a shot selection would be very difficult to grab in the preview in List view. Instead, let's switch to Filmstrip with the List view button or use the shortcut **Option-Command-1**.

2. Depending on the size of your monitor, adjust the thumbnail slider to either 5 or 10 seconds per thumbnail.

3. Go down to 31:09, mark an In point, and mark the Out point at 34:15.

4. Grab the selection directly from the Event Browser and drag it to the Timeline, right at the beginning, so it pushes everything out of the way, as in Figure 5.6. This is an Insert edit; it inserts the clip between other clips in the Timeline.

5. Let's add another piece of the same shot to the Timeline, but first let's position the playhead to where we want to go. Either drag the playhead to the edit point between the first and second shots, or move the skimmer to that point and click to bring the playhead there.

FIGURE 5.6

Inserting a Clip into the Timeline.

FIGURE 5.7

Snapping Button.

As you move the playhead or skimmer onto the edit point, it should snap strongly to the join. If it doesn't, that means **Snapping** is turned off. Press the **N** key or click the Snapping button in the upper right of the Timeline (see Figure 5.7). Blue means it's on, and gray means it's off. Try it several times, toggling the Snapping function on and off. The **N** key may become the one you use the most in Final Cut Pro. You'll probably be changing often from one mode to the other. Now with Snapping on, you should have the playhead parked precisely between the clips.

NOTE

Insert into a Clip

Unlike in earlier versions of FCP, in this version you cannot insert into the middle of a clip by dragging onto a clip in the Timeline. Doing so will produce a Replace edit. To achieve an insert anywhere, use the menu command or the keyboard shortcut **W**.

With the playhead between the two clips, notice the slightly translucent L in the lower-right corner of the Viewer (see Figure 5.8). This overlay indicates that you're on the first frame of that clip in the Timeline. If you step back one frame, you'll see a reverse L in the lower-right corner of the Viewer, indicating you're on the last frame of that clip in the Timeline (see Figure 5.9). Make sure the playhead is back at the edit point so you see the first frame indicator in the Viewer.

FIGURE 5.8

First Frame Overlay.

FIGURE 5.9

Last Frame Overlay.

TIP

Fit Content to the Window

One of the most useful keyboard shortcuts has been carried through since the earliest versions of FCP into the newest version: the shortcut to zoom to fit as it's now called. The shortcut is **Shift-Z**, and it works to fit all the clips in the project into the available space of the Timeline panel. It also works to fit the content of the Viewer into its available space. I find I use it a lot.

TIP

Moving Between Edit Points

The simplest way to move between edit points is to use the **Up** and **Down** arrow keys. The **Up** arrow takes you back one edit event. It moves the playhead back from where you are to the edit before. The **Down** arrow key moves the playhead to the next edit event, whether an audio or video edit point, whichever comes next in the Timeline. You can also use the **semicolon** key to move backward through the edits and the **apostrophe** key to move farther down the Timeline. Another handy tool, especially for playback and review, is Editmote from Digital Rebellion (www.digitalrebellion.com/editmote/), which is an iOS-based iPhone or iPod Touch remote control for FCP. It also works with a number of other applications such as After Effects and the QuickTime player. It requires that both workstation and remote be on the same Wi-Fi network.

Let's insert a shot between edit points:

1. With the playhead positioned where you want to insert the next shot, between the first and second shots in the project, let's make another selection from the *cutting meat* clip, a close-up section where the meat is being cut.
2. Go to 26:06 and mark an In point for the start of the selection. Mark an Out point at 29:06.
3. From the **Edit** menu, select **Insert**.

Immediately, the Timeline rearranges itself. The second piece of *cutting meat* drops into the Timeline between the first and *man standing at counter*, pushing everything farther down in the Timeline.

TIP

Going to a Place in the Video

To go to a specific frame of video, either click the timecode display in the Dashboard, or press **Control-P** for the playhead (see Figure 5.10) and type in the number you want, such as *2606*, and press **Enter**. You don't need to type in the colon or semicolon.

FIGURE 5.10

Moving the Playhead in the Dashboard.

FIGURE 5.11

Insert Button.

Let's do a couple more Insert edits:

1. Let's go to the close-up of the butcher, starting at 20:06 and ending at 22:10.
2. To execute the Insert edit, click the **Insert** button in the Toolbar (see Figure 5.11). If the playhead in the Timeline is at the end of the second *cutting meat* shot, that's where the clip will be inserted into the project.
3. We can get a fourth shot out of this long clip, starting at 55:22 and ending at 58:14. It's the shot of the man watching the butcher.
4. To edit this into the Timeline after the other *cutting meat* clips, use the keyboard shortcut **W** to insert the shot. The shortcuts are the best ways to do these types of edits in my view.

Connect

A **Connect** edit is something new to Final Cut. In this version, clips are either on the primary storyline or connected to the primary storyline. Everything connects to this one flow of material and time that runs through the center of the project. Connected Clips are not new to iMovie users, but they are much more flexible and hold more opportunities in FCP because they can be converted into compound clips and additional storylines, which we'll see later. Let's start by making a Connected Clip, putting a cutaway in the middle of the *man handed bag* shot.

1. Move the Timeline playhead to around 19:20, a beat after the camera pulls back and the image brightens. This is where we'll connect the first clip.
2. Select the *prosciutto samples* clip and mark an Out point at about 3:00.
3. Before we edit this into the Timeline, let's switch off the audio, so only the video is put into the Timeline. To do this, use the little triangle menu next to the edit buttons in the Toolbar. Select **Video Only** or use the keyboard shortcut **Option-2**. In Figure 5.12, you'll see the way the buttons change, showing blue icons indicating Video Only edits.
4. You could use the **Edit>Connect to Storyline** function, but instead use the Connect button in the Toolbar, the first of the three edit buttons (see Figure 5.13).

FIGURE 5.12

Video Only Edits.

FIGURE 5.13

The Connected Clip Button.

The clip appears in the Timeline with a little connecting pin hanging from it (see Figure 5.14); this pin connects the start of the clip to a specific frame of video on the primary storyline. You can easily reposition and move Connected Clips by dragging left and right along the Timeline as needed.

We did a Video Only edit, but you can also do an Audio Only edit, also selected from the edit button pop-up menu or with the shortcut **Option-3**. The buttons will then be displayed as in Figure 5.15. The keyboard shortcut **Option-1** will return to doing All, both video and audio together. These selections effect not only the edit buttons, but also other edit functions, such as Append, Inserts, Connects, or Overwrites, which are performed with the **QWED** keyboard shortcuts.

Let's add part of the *dried tomatoes* clip right after the first Connected Clip, but let's use a slightly different technique, defining the duration of the clip in the Timeline:

1. With the playhead at the end of the first Connected Clip, mark the beginning of a selection with the **I** key. This marks the clip on the primary storyline from that point to the end of the clip.
2. Play forward till the woman in the foreground clears the frame, about 25:16, and mark the end of the selection with the **O** key. My selection is just under three seconds.

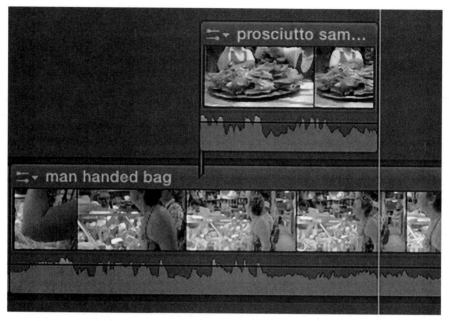

FIGURE 5.14

The Connected Clip in the Timeline.

FIGURE 5.15

Audio Only Edits.

3. Click on the *dried tomatoes* clip in the Event Browser to select it. The whole clip, which is 17 seconds long, is selected.

4. Press the **Q** key to execute the Connect edit. The clip takes its duration from the selection in the Timeline.

This technique allows you to edit with great precision, but let me show you another useful trick:

1. Undo the last edit with **Command-Z** so that the Connected Clip disappears from the Timeline, but the Timeline selection remains.

2. Play the *dried tomatoes* clip until almost the end after the woman arranging the olive oil leaves the frame around 15:29 and mark an Out point. The selection is from the beginning of the clip to the Out point and is 16 seconds long.

3. Instead of simply connecting the clip, we're going to backtime the connection. Press **Shift-Q**.

The shot is connected to the Timeline for the same duration, using the marked Out points as the reference, the Out point in the browser matched to the Out point in the storyline, and the duration backtimed from that point. Other edit functions such as Insert and Overwrite can use this technique as well.

Connected Clips are most commonly used to add B-roll shots to an interview or a narration. Connected Clips are also used to add clips to a music video, where the music track itself is placed on the primary storyline and the pictured connected to it.

What Is B-Roll?

You've seen B-roll, even if you do not know the term or what it means. The A-roll is the interview or narration or primary audio that's used in the piece. The B-roll is the video that overlays this to illustrate what the person is talking about: You don't see the person; you see what he's talking about. The term comes from the days when television news was on film. Two film chains would roll simultaneously. One machine played back the A-roll, which contained the primary interviews and the reporter on camera, while the other machine played the B-roll. The director could then switch back and forth between the two film chains as needed. Often the B-roll chain was silent, and all the audio came from the A-roll, though sometimes the B-roll audio was mixed in or played at full volume.

Overwrite

Though there is also an Overwrite edit function in FCP, it is not as widely used in the current version as in previous versions. For iMovie and legacy Final Cut users, an Overwrite edit is one in which the video will punch into the storyline overwriting what's already there, cutting it out for the duration of the new shot. There is no button for the Overwrite edit, and you cannot overwrite by dragging into the Timeline with the Select tool, but you can use the menus or a keyboard shortcut.

1. As before, we'll start by marking a selection in the Timeline. Mark an In point with the **I** key at the beginning of *ls woman getting sample*, about 30:24 in the Timeline, and mark an Out point at 33:02, about two seconds and nine frames later.

2. Press **Option-1** to reset the edit functions to All so that you can edit both video and audio into the Timeline.

3. Select the *dried tomatoes* clip again, and this time we'll use the beginning of the shot. With the clip selected, use **Edit>Overwrite** or the shortcut, which is the **D** key.

The clip punches into the storyline, cutting out what's there, replacing it with the shot from the browser, taking its duration from the selection in the storyline (see Figure 5.16).

Another useful Overwrite function is the ability to collapse Connected Clips in the primary storyline.

1. Select the two Connected Clips in your sequence.
2. You can then either use **Edit>Overwrite to Primary Storyline** or the keyboard shortcut **Option-Command-Down** arrow to put the two clips into the primary storyline. Your Timeline will then look like Figure 5.17.

Notice the audio for the adjacent clips expands out to make room for the audio from the new clips and is not wiped out when Connected Clips use the Overwrite to Primary Storyline function. We'll look at expanding clips in Lesson 7 on audio.

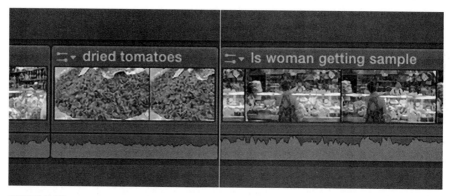

FIGURE 5.16

The Timeline After Being Overwritten.

FIGURE 5.17

Connected Clips Overwritten into Primary Storyline.

You can also lift clips out of the primary storyline and make them Connected Clips.

1. Select the two clips you just collapsed and use either **Edit>Lift from Storyline** or press the keyboard shortcut **Option-Command-Up** arrow. The two clips lift from the storyline and appear as clips connected to two Gap Clips on primary storyline (see Figure 5.18).
2. Press **Command-Z** twice to undo the lift and to undo the overwrite into the storyline. The audio of *man handed bag* remains collapsed.
3. Select the *man handed bag* clip and you can either use **Clip>Collapse Audio/ Video,** or you can use the shortcut **Control-S** or simply double-click the audio to collapse it back into the video.

Table 5.1 shows the primary editing keyboard shortcuts we've used to add clips to the project.

Replace

The **Replace** edit function is very different from the Replace function in previous versions of Final Cut Pro, and although different, it brings some very powerful tools to the application. Replace is remarkably sophisticated. It will replace a clip in the

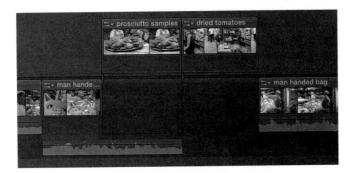

FIGURE 5.18

Primary Storyline Clips Converted to Connected Clips.

Table 5.1 Principal FCP Edit Shortcuts	
Edits	**Shortcut**
Connect	Q
Insert	W
Append	E
Overwrite	D
All (Video and Audio Edit)	Option-1
Video Only Edit	Option-2
Audio Only Edit	Option-3

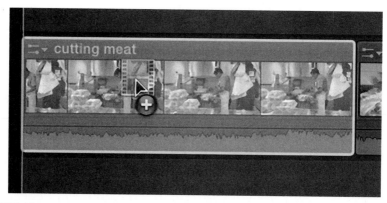

FIGURE 5.19

Dragging to Replace.

Timeline with another clip from the Browser. It replaces it in a number of different ways: a full replace ignoring duration, a replace timed forward from the start of the clip, a replace backtimed from the end of the clip, and a replace with Audition—a great new feature for procrastinators and the indecisive. Let's do a Replace edit:

1. Start with the first shot in the Timeline, the long shot portion of *cutting meat,* which is 3:07 in length. In the Event Browser, find *ls cleaning fish,* which is eight and a half seconds.
2. Drag *ls cleaning fish* onto the first *cutting meat* shot; hold till the Timeline clip turns white, as in Figure 5.19; then release the clip to bring up the shortcut menu.
3. Select **Replace**, and the whole clip drops into the Timeline, replacing the one that was there and pushing everything else out of the way.
4. Undo that replace with **Command-Z.**
5. Repeat the process dragging the *ls cleaning fish* clip to the replace, but this time from the shortcut menu that appears, select **Replace from Start**. Now the clip takes its duration from the clip that was already in the Timeline, starting from whatever In point you had on the browser clip.
6. Select the second *cutting meat* shot in the Timeline, and in the Event Browser, find the *cleaning fish* clip.
7. Play the clip till the camera settles on the medium-long or three-quarter shot around 22:27 and mark an In point. The duration is 9:18.
8. With the second *cutting meat* clip selected in the Timeline, press **Shift-R**. This will do the full Replace edit, replacing the duration of the Timeline clip with the duration of the clip in the browser.
9. Again, undo that with **Command-Z.**
10. Instead of the full replace, press **Option-R** for the **Replace from Start** edit function.

11. Let's replace the third *cutting meat* with a section of *cleaning fish*. Select the *cutting meat* clip in the Timeline and select the whole *cleaning fish*.
12. On the *cleaning fish* clip, mark an Out point at 14:25 with the **O** key.
13. Drag the selection from the browser and hold it onto the Timeline clip so it goes white. From the shortcut menu, select **Replace from End**. The close-up shot will replace the close-up of the butcher that was in the Timeline.

This is a really handy and efficient way to change shots and try different material, but there is an even more powerful tool in the shortcut menu, which we'll see next.

TIP

Multiple Clips

You aren't limited to dragging a single clip to the Replace function. If you wish, you can select two or more clips and drag them into the Timeline to replace a single clip.

Audition

Audition is a unique Replace function. It lets you create a container of clips that you can easily switch between to replace a clip in the Timeline to try many different options. To see what it does and how it works, we need to do a Replace edit:

1. From the *cleaning fish from behind* clip, select a short portion of about three seconds in the close-up of cutting the fish. Mark an In point at 4:00.
2. To make the selection three seconds, you can either double-click the timecode in the Dashboard or press **Control-D** for duration.
3. Type in *3.* (that's three and period) or *300,* as in Figure 5.20, and press **Enter**. The clip's duration will be set to three seconds.
4. We're going to replace the last *cutting meat* shot, so drag the selection from the Event Browser to the Timeline onto the remaining *cutting meat* clip, hold for it to turn white, and then select **Replace and add to Audition**.

If you zoom into the Timeline you'll see that the clip has a badge in the upper-left corner that looks like a spotlight. Click on it to open the Audition HUD (see Figure 5.21). In the HUD, you see both clips: the current clip in the Timeline, plus the clip it replaced. To select the other clip, simply click on it to change the clip in the Timeline. You can also use **Left** or **Right** arrow to switch between the two.

FIGURE 5.20

Setting the Duration in the Dashboard.

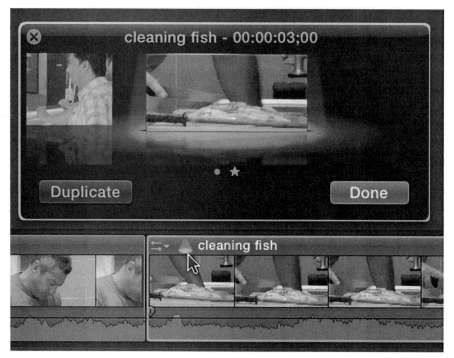

FIGURE 5.21

Audition HUD.

An interesting feature is the behavior of the spacebar play function while the HUD is open. If you press the spacebar, the playhead will start playing the Timeline before the clip, play through the clip, and past the clip. It will automatically be in Looped Playback mode, so it just keeps playing over and over while you try different possible selections. Do you want the close-up of the fish being cut, or do you want the man standing and apparently watching? You can add as many clips to the Audition container as you want and keep switching between them.

Notice the HUD also has the option to duplicate a clip. This allows you to have multiple copies of the same clip with different effects applied to them, perhaps with different color looks. We'll see this in Lessons 10 and 11 on effects and color correction.

With the clip selected in the Timeline, you can open the Audition HUD either from the **Clip** menu using the **Audition** submenu or simply by pressing the **Y** key. Another really useful shortcut is **Control-Command-Y**, which not only opens the Audition HUD, but also sets off playback, putting the playhead in Looped Playback mode.

Though you can simply leave the Audition container in the Timeline throughout your project, if you do finally decide on a specific shot, you can select **Finalize Audition** from the **Clip>Audition** submenu or use **Option-Shift-Y** to close the container and leave the final shot in the Timeline.

SUMMARY

In this lesson you learned how to use the editing tools that move clips into the Timeline: the Append edit, to add the end of the Timeline; the Insert edit, to push a clip into the storyline to make room for itself; the Connect edit, to attach a clip to the primary storyline, the Overwrite edit, to push into the storyline; and the Replace edit, to change a shot in the Timeline. You used drag and drop, the menus, the buttons, and the keyboard shortcuts. You also saw how to edit video and audio independently, and to use Audition as a replace function to try out different clips. Now that we've looked at how to edit shots into the Timeline, we'll next see how to adjust and trim and reposition clips in the project.

Advanced Editing: Trimming

6

Now that you've cut up your material in Final Cut Pro and learned the editing functions, it's time to fine-tune your project. There is no right or wrong way to edit a scene or a sequence or even a whole film or video; there are only bad ways, good ways, and better ways. Trimming in FCP is done in the Timeline, using selections and special tools.

LOADING THE LESSON

For this lesson we'll use the same *FCP3.sparseimage* as we used in the preceding lesson. If you haven't downloaded the material, or the project or event has become messed up, download it again from the www.fcpxbook.com website. Then do this:

1. Once the file is do wnloaded, copy or move it to the top level of your dedicated media drive.

2. Double-click the *FCP3.sparseimage* to mount the disk.
3. Launch FCP.

The media files that accompany this book are heavily compressed H.264 files. Playing back this media requires a fast computer. If you have difficulty playing back the media, you should transcode it to proxy media:

1. To convert all the media, select the event in the Event Library, **Control**-click on it, and choose **Transcode Media**.
2. In the dialog that appears, check on **Create proxy media**.
3. Next, you have to switch to proxy playback. Go to **Preferences** in the **Final Cut Pro** menu.
4. In the **Playback** tab, click on **Use proxy media**. Be warned: If you have not created the proxy media, all the files will appear as offline, so you will have to switch back to **Use original or optimized media**.

The sparse disk has two projects in its Project Library: one called *Trim* and the other called *Trim2*. We'll be working with these projects in this lesson, together with the event called *Edit* in the Event Library. Inside the event, you'll see the media we're going to work with. Before we begin, I suggest duplicating the *Trim* project:

1. Select the project listed under the *FCP3* disk icon and use **Command-D** to duplicate it.
2. Leave **Duplicate Project Only** checked and uncheck **Include Render Files**.
3. Duplicate the project to your system drive or your media drive outside the *FCP3* disk. This will preserve the original in the *sparseimage* disk for you to return to if necessary.
4. You can rename it to something like *Trim copy*.
5. Open the duplicate project.

TRIMMING SHOTS

FCP has several tools for refining your edit, fine-tuning, and rearranging the sequence. Let's look at some of those, working with the shots that we have edited together into the Timeline.

Moving Shots

Very often you'll want to rearrange shots, changing their order in the sequence. In earlier versions, this was called a *shuffle* edit or a *swap* edit. Unlike earlier versions, where we could shuffle only one shot at a time without copying and pasting, in this version, we can move as few or as many shots as we want. We're going to move everything from after *man standing at counter* until the end of the project, to the beginning of the sequence:

FIGURE 6.1

Shuffle Edit in the Timeline.

1. You can drag-select the clips, or you can select *man standing at counter* and then, holding the **Shift** key, select the last clip *ls woman getting sample.*
2. Drag the clips to the front of the Timeline until everything else pushes out of the way and drop the clips, as in Figure 6.1.

You'll notice that not only have the clips on the primary storyline been moved, but the Connected Clips attached to them have moved as well. Everything moved in one large piece, inserted itself back into the project, shuffling everything else out of the way.

You can also do this by copying and pasting:

1. Undo the previous Shuffle edit with **Command-Z**.
2. Select *man at counter* and all the clips after it. Be careful: You MUST select the Connected Clips as well. If you don't, the Connected Clips will disappear. This is only necessary when doing copy and paste.
3. Use **Command-X** to cut the clips from the Timeline.
4. Move the playhead to the beginning of the sequence with the Home key and press **Command-V** to insert the cut clips back into the project at the start.

Everything after that point is pushed out of the way to make room for the clips. The paste function usually does an Insert edit, unless you're copying and pasting items that aren't on the primary storyline.

Changing Attachment

Sometimes you want to move Connected Clips, and when you do, the attachment points, which are at the head of the clip, move as well, perhaps to another clip that you don't want them attached to. You can change the attachment point fairly easily with some precise clicking. Let's start by moving the two Connected Clips a little earlier in the Timeline. We'll ignore how they look as edits.

1. Grab the two Connected Clips, *prosciutto samples* and *dried tomatoes*, and drag them to the left so they bridge the first edit, as in Figure 6.2. The first Connected Clip will be connected to the first clip, and the second Connected Clip to the second in the primary storyline. We want the first Connected Clip attached to the second as well.

FIGURE 6.2

Moving Connected Clips.

FIGURE 6.3

Reattaching Clip.

2. Holding the **Option** and **Command** keys together, click on the *prosciutto samples* clip above the second clip on the primary storyline, *man handed bag*, as in Figure 6.3. The attachment point will jump to your click point.
3. Use **Command-Z** to undo the move of the two Connected Clips.

Adding Edits and Gap Clips

Let's look at some simple ways to cut clips that are in the Timeline. To see this, let's append a shot into the Timeline:

1. Select the *cleaning fish from behind* clip and press **E** to append the whole clip into the end of the project.

FIGURE 6.4

Tools Menu.

2. Play the *cleaning fish from behind* clip in the Timeline until the camera pushes in and settles at 54:24 in the Timeline.

3. Make sure Snapping is turned on, and from the Tools pop-up, select the Blade tool (see Figure 6.4) or press **B** to activate it.

4. Drag the pointer across the clip in the Timeline till it snaps to the playhead. Notice that when the Blade tool is over a clip, the Dashboard no longer shows the Timeline time, but the elapsed time of the clip underneath the Blade tool (see Figure 6.5).

5. Click with the Blade tool at the playhead to cut the clip. This is an Add edit. It adds an edit point to the clip.

6. Undo that edit to see a variation of using the Blade tool that can be very helpful.

7. Press the **A** key (A for arrow) to return to the normal Select tool.

8. Move the pointer so it snaps to the playhead; then hold down the **B** key before clicking the clip. When you release the **B** key, the pointer reverts to the Select tool. Most of the tools have this very useful capability.

9. Select the cut portion at the head of the clip and press the **Delete** key. On most keyboards, this is the large **Delete** key next to the equal (=) key. In FCP, this will do a Ripple Delete or Extract edit, removing the section from the Timeline and closing the gap. This **Delete** key should not be confused with the **Forward Delete** key on an extended keyboard, which performs a different function.

FIGURE 6.5

Dashboard and Cut Clip in the Timeline.

FIGURE 6.6

Gap Clip.

10. If you want to do a Lift edit, which removes the clip from the Timeline but leaves the gap, press **Shift-Delete** or the **Forward Delete** key on an extended keyboard. This will leave a Gap Clip in the Timeline in place of the deleted clip, as in Figure 6.6.

Gap Clips are objects in the Timeline. They are similar to slugs in legacy Final Cut, except they are transparent and have no audio. Gap Clips can be cut, edited, trimmed, and perhaps most importantly, have Connected Clips attached to them.

They make great placeholders for missing interview shots where supporting B-roll can be connected to the Gap Clips.

Let's look at another way to make an Add Edit:

1. First, undo the edits you did until you're back to the complete *cleaning fish from behind*. Use **Command-Z**. If you step back too far, use **Shift-Command-Z** to redo.
2. Put the Timeline playhead at the same position as before, 54:24.
3. Rather than selecting the Blade tool, simply press **Command-B** to make an Add edit at the playhead, cutting the clip in the Timeline.
4. Select the cut at the beginning of the clip and press **Delete** to remove that section.

TIP

Moving the Playhead

Remember the **Left** or **Right** arrows move the playhead in one-frame increments and **Shift-Left** or **Shift-Right** arrow will move the playhead forward or backward in 10-frame increments.

TIP

Multiple Tracks

Using **Command-B** is the best way in FCP to cut through multiple tracks of video and audio that you need to delete. You have to select the items on the tracks you want to cut, either by drag-selecting, **Command**-selecting, or by pressing **Command-A** (Select All) with the playhead parked at the desired edit point and then cutting them with the shortcut.

Marking a Selection

Instead of using the Add Edit function with **Command-B**, you can remove sections by marking an In and Out point in the Timeline. One of the important features of marking a selection or using the Range Selection tool is that it can bridge across an edit or multiple edit points. Now try this:

1. Go back in the Timeline to the two shots called *ls woman getting sample*.
2. Play in the Timeline until about 24:11, just before the camera starts to move forward.
3. Press **I** to mark an In point in the Timeline, which is marked on the primary storyline clip, from that point until the end of the clip.
4. Play forward until 30:27, across the edit point, into the second *ls woman getting sample,* until just before the woman starts to take off her glasses.
5. Press **O** to mark the end of the selection. (You can then use the **Delete** key to remove that section, but don't do that now so that I can show you a couple more methods first.)
6. Press **Shift-Command-A** to deselect all and drop the selection.

> **NOTE**
>
> Marking the Timeline Ruler
>
> Unlike in previous versions of Final Cut, in the current version you cannot mark In and Out points on the Timeline Ruler, only on clips themselves. That said, marking these selections very often behaves the same way as in previous versions. Also, you cannot apply markers to the Timeline Ruler, only to clips. This means that you cannot create chapter markers for DVDs in the new version of FCP. Chapter markers can be created only in Compressor or the DVD authoring application.

Range Selection

Another way to mark a selection would be to use the Range Selection tool that can be called up from the Tools menu in the Toolbar or with the **R** key. Like the Blade, it can also be called up temporarily by holding down the **R** key. Let's make the selection that way:

1. Because of the lack of timecode display when dragging with tools in the Timeline, it's easiest to move the playhead to the end position so your selection can snap to it. Put the playhead at 30:27.
2. Move the skimmer across the clip in the Timeline until you get to 24:11. Then hold down the **R** key and drag across the clips till the tooltip shows 6:15 and snaps to the playhead, as in Figure 6.7.
3. Then release the mouse and release the **R** key. The selection is made on the clips, across the edit point.
4. Press the **Delete** key to remove the section.

Trim Shortcuts

The **Trim Shortcuts** are new to this version of FCP and are a wonderful advancement, making trimming in the Timeline very powerful.

Let's begin by trimming some more from the start of the *ls woman getting sample* clip, using the new edit function called **Trim Start**:

1. Play the start of the clip in the Timeline until you're at about 27:09, shortly after she takes off her glasses.

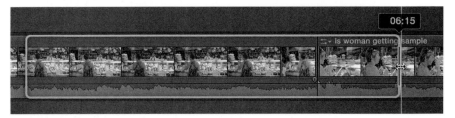

FIGURE 6.7

Range Selection in the Timeline.

2. Select **Trim Start** from the **Edit** menu or, even easier, press **Option-[** (that's the left bracket key). That removes everything from the start of the clip up to that point, up to where the playhead is.
3. Let's take some off the end of the first shot, *man standing at counter*. Play the Timeline from the start for about three and a half seconds, and use **Trim End** from the **Edit** menu or press **Option-]** (that's the right bracket) to trim from the playhead to the end of the clip.

Let's do a **Trim to Selection** edit, which allows us to keep a section in the middle of the clip and will "top and tail the shot," as it's called. For this, we'll use the *cleaning fish from behind* shot at the end of the sequence. You can mark the selection of the area you want to keep either with the Range Selection tool or by marking In and Out points for greater precision. Then do this:

1. Play the clip until about 48:00, until just after the camera pulls back and stops.
2. Press **I** to mark the beginning of the selection.
3. Play forward until just before the camera starts to push in, around 53:12, and mark the end of the selection with the **O** key.
4. With the selection made, use **Edit>Trim to Selection** or the keyboard shortcut **Option-** (the backslash key) to trim the top and tail of the clip, leaving only the center selected section in the Timeline.

TIP

Topping and Tailing on the Fly

If you're feeling very confident or if you just need to roughly trim shots while playing back, you can simply press the shortcuts **Option-[** or] on the fly as the project plays. No clips have to be selected. If the only clips are on the primary storyline, the trim functions will trim those clips. If there are Connected Clips above, the trim function will top or tail the shots on top, not the shots on the primary storyline. However, if the clips on the primary storyline are selected, then the trim start and trim end functions will cut the primary storyline clips, not the Connected Clips above them.

TRIM TOOLS

Whenever you shorten or lengthen a clip in FCP to fine-tune your edit, the application will always want to alter or change the length of the sequence, making it shorter or longer as needed. This is FCP's Magnetic Timeline. If you do not want the magnetism of FCP's Timeline, you can override it with the Position tool as we will see.

There are basically four types of trim edits: **Ripple, Roll, Slip,** and **Slide**. Where there used to be separate tools for each of the trim functions, in FCP there is now only a single Trim tool; the other functions are either available automatically or are incorporated using the single tool. Each trim edit works differently: the Ripple and Roll edits change the duration of clips, while the Slip and Slide edits leave the clip duration intact.

A *Ripple* edit moves an edit point up and down the project by pushing or pulling all of the material on the storyline, shortening or lengthening the whole sequence. In a Ripple edit, only one clip changes duration, getting longer or shorter. Everything else that comes after it in the track adjusts to accommodate it.

A *Roll* edit moves an edit point up and down the project between two adjacent shots. Only those two shots have their durations changed. One gets longer, and the adjacent shot gets shorter to accommodate it. The overall length of the project remains unchanged.

A *Slip* edit changes the In and Out points of a single clip. The duration of the clip remains the same, and all of the clips around it remain the same. Only the selected content of the slipped clip changes. If more frames are added on the front, the same number of frames is cut off the end, and vice versa. If some are added to the end, an equal amount is removed from the beginning. In a Slip edit, neither its position in the project nor either of the adjacent shots is affected.

A *Slide* edit moves a clip forward or backward along the project. The clip itself, its duration, and its content remain unchanged. Only its position on the project, earlier or later, shortens and lengthens the adjacent shots as it slides up and down the track. The shot you're sliding doesn't change; only the two adjacent shots change.

You can trim an edit by:

* Dragging
* Nudging
* Numerically extending

The Ripple Edit

Let's begin by duplicating the project in the *FCP3* disk image called *Trim2*. This will give you a clean project to work on and keep the original to go back to if you need to. Switch to the Project Library (**Command-zero**) and duplicate the project from the **File** menu or with the shortcut **Command-D**, just as in the Finder, and save the duplicate to your system drive or your media drive, not in the *FCP3* disk image. So that you don't confuse the two projects, give the duplicate a new name.

To trim an edit, you have to select an edit. You can do this by clicking on an edit point or by using a keyboard shortcut:

1. Move the pointer to the left side of the edit point between the first and second shots, *fresh tomatoes* and *cheese*. The pointer will change to the Ripple tool, as in Figure 6.8.
2. If you move to the right side of the edit point, the pointer will change to ripple the other side of edit, as in Figure 6.9.

    This is a single-sided edit. It will change the clip on only one side of the edit point at a time—either the outgoing side, the A side, when the tool is on the left side of the edit; or the incoming side, the B side, when the tool is on the right side of the edit. Click on either side of the edit point to select the edit.
3. Click and drag on the left side of the edit. A time value will display above the edit as you drag. Pulling to the left will make *fresh tomatoes* shorter, and pushing to the right will make *fresh tomatoes* longer, as in Figure 6.10.

FIGURE 6.8

Ripple Tool in the Timeline.

FIGURE 6.9

Ripple Tool on the B Side.

FIGURE 6.10

Rippling a Clip to Make It Longer.

> **TIP**
>
> Avoid the Audio Fader
>
> Be careful where you're dragging to ripple because you can also drag the audio fade if your cursor is unintentionally parked on it. If a double-headed arrow appears, it tells you this is *not* the Ripple function.

As you shorten or lengthen the clip, the rest of the project adjusts itself, getting longer or shorter as needed. Everything to the right of the edit point moves in the Timeline. If you have **Show detailed trimming feedback** switched on in the FCP preferences, which you should always do, you'll see the edit point displayed in the Viewer (see Figure 6.11). The image on the left will show the outgoing shot, which will change as you drag the outgoing side, and the image on the right will show the first frame of the incoming shot, which will remain unchanged as you drag on the left side.

Notice that when you select the edit point, it turns yellow, indicating it's selected and ready to be trimmed. Click on the right side of the edit point at the start of *prosciutto*. Notice it's red rather than yellow. This indicates there is no more available media to lengthen the shot. All of the shot is in the Timeline. You can't make the shot longer, but you can, of course, make it shorter.

You can select an edit point to trim it either by clicking on it, as we just did, or by moving the playhead to the edit point and pressing either of the **bracket** keys: the **left bracket** ([) to select the outgoing side or the **right bracket** (]) to select the incoming side.

With the edit selected in Trim mode, you can nudge the clip earlier or later in time using the **comma** (earlier) or **period** (later) to nudge the edit one frame at a time. **Shift-comma** or **Shift-period** will move the edit in 10-frame increments.

FIGURE 6.11

Detailed Trimming Feedback.

FIGURE 6.12

Media Limit Indicator.

FIGURE 6.13

Numerical Ripple.

To trim the clips numerically, you simply have to type + or − and then a time value to move the edit later or earlier in the Timeline. Let's trim the *prosciutto* clip:

1. To select the right side of the edit at the start of the clip, move the playhead to the edit point and press the **right bracket** key. The selection will be red, as in Figure 6.12 (also in the color section).
2. To move the edit later in time (you can't move it earlier because there isn't any more media), type **+1.** (plus one and period) for one second. The Dashboard will display the value, as in Figure 6.13.
3. Press **Enter** to execute the edit.

Unlike in earlier versions of FCP, in this version all the trim tools can be changed using an **Extend** edit function, including Ripple and Roll edits. To do an Extend edit, you move the playhead to where you want the edit to move to. You can move the playhead either after the Trim edit is selected or before. If you move the playhead first, you'll have to be careful how you select the edit. Let's move the playhead first:

1. Move the playhead to 3:15 in the Timeline. You can either click in the Dashboard, or you can press **Control-P** and type in **315** and press **Enter**, or just drag the playhead to the right position.
2. If you click on the outgoing edge (the left side) of the edit point, the playhead will move there. However, if you hold the **Option** key and click the edit point to select it, the playhead will remain where you placed it.
3. With the edit selected, use **Edit> Extend Edit** or the keyboard shortcut **Shift-X** to ripple the edit to the playhead.

These functions used to adjust edits using the Ripple tool apply to all the other edit tools as we will see. The only difference is the way in which the edit or clip is selected. This uniformity of behavior is one of the best features of trimming in FCP.

The Roll Edit

Unlike the Ripple edits, which are automatically accessed when the Select tool moves to the edit point, the Roll edit needs to use a special Trim tool if you want to drag the edit. The tool can be accessed from the Tools menu in the Toolbar or by pressing the **T** key.

The Roll edit does not affect the overall length of the sequence, only the two adjacent shots, making one longer and the other shorter. Let's roll the edit between *ms ham* and *cu ham:*

1. Turn on the Trim tool with the **T** key.
2. Click the edit point to select it. Both sides of the edit are selected.
3. Drag the edit point to the left to shorten the outgoing A side and lengthen the incoming B side, as in Figure 6.14.

    If you drag it –27 frames as I did, you'll see that the selection indicator on the incoming side turns red, showing that you are on the first frame of the incoming media, and that's as far as the clip can be dragged. Also notice the filmstrip overlay on the left edge of the incoming shot in the Viewer; it indicates you're at the limit of the media (see Figure 6.15).

    You can also do this edit numerically and by nudging. To do it numerically or to nudge it, you do not have to have the Trim tool selected; you simply have to select the edit in Roll mode.
4. Undo the previous edit with **Command-Z**.
5. Deselect the edit by clicking off it or pressing **Shift-Command-A** to deselect all.
6. Return to the Select tool by pressing **A** or by choosing it from the Tools menu in the Toolbar.

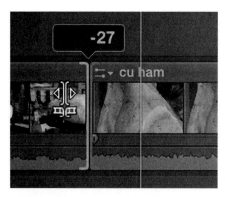

FIGURE 6.14

Rolling an Edit in the Timeline.

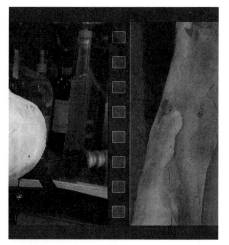

FIGURE 6.15

Media Limit Indicator in the Viewer.

7. Make sure the playhead is on the edit between *ms ham* and *cu ham* and press the **backslash** (\) key to select the edit in Roll mode.
8. Type in −29 and press **Enter**. The edit will roll earlier in time but will go only as far as there is available media, showing the red media limit indicator on the right side of the edit point.

You can nudge the edit in the same way, select it, and use the **comma** or **period** keys, or **Shift-comma** or **Shift-period**. You can also do an Extend edit by moving the playhead to the point where you want the edit to go and pressing **Shift-X**. Unlike in earlier versions of Final Cut, including FCE, in this version if there is not enough media for the edit, rather than not doing the Extend edit at all, FCP will now move the edit point as far as the given available media. In this case it will move it only the maximum 27 frames as we just did.

TIP

Temporarily Using the Trim Tool

Rather than tapping the **T** key to activate the Trim tool, if you hold the **T** key to roll an edit point, when you release the **T**, the pointer will return to the Select tool. This trick works well for making fast adjustments to your edits.

The Slip Edit

Slip edits move the content of the shot without changing the clip's duration or its place in the Timeline. This is one of the most useful Trim tools for making small adjustments to clips in the project. To use it, you have to activate the Trim tool. No separate slip or slide tools are needed in FCP, unlike earlier versions. The same tool used to roll an edit is used to slip the contents of a clip.

Let's look at the second shot in the project *cheese*. If you play the shot, you'll see it pans from right to left across a cheese display. Rather than the pan, perhaps you'd like to use the static portion at the beginning of the shot or maybe the portion near the end of the shot. To do this, you do a Slip edit:

1. Select the Trim tool from the menu or press **T** for trim.
2. Click on the *cheese* shot to select it. The icon will change to the Slip tool, and the two ends of the clip will be selected, bracketed inward (see Figure 6.16).
3. Drag the clip as far to the right as you can. As you do, you'll see the contents slip in the Timeline filmstrip and also in the two-up display in the Viewer.

By moving to the right, you are selecting a piece of the clip that's earlier in the shot. Now the pan has been eliminated, and the shot is mostly static. If you want to nudge the shot, select it in Slip mode and use the **comma** or **period** keys as we've done before. To slip it numerically, select the clip in Slip mode and type in a − or + time value. The clip will slip that amount or as far it can assuming available media.

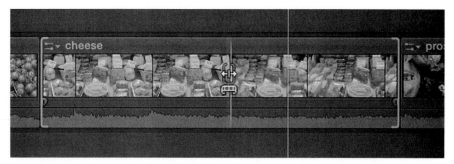

FIGURE 6.16

Clip Selected for Slip Edit.

> **NOTE**
>
> Extending in Slip Mode
>
> Though you can do an Extend edit to slip earlier in time, with the playhead to the left of the clip, you cannot do an Extend edit to slip the contents later in time, with the playhead after the clip. It will simply wipe the shot out when you press the keyboard shortcut, leaving just one frame in the Timeline, which is not a good thing. Basically, you should forget using the Extend edit function with the Slip and Slide functions.

The Slide Edit

Let's look at the last trimming function, the Slide edit. This leaves the clip you're sliding untouched, simply moving it earlier or later in the project and making the adjacent clips longer or shorter as needed. Though this can be a useful tool in some situations, I find in practice that I don't use it very often; most commonly I use it when editing music. Let's see how to do it using the *cheese* shot in the Timeline:

1. Make the Trim tool active by pressing the **T** key.
2. Instead of simply clicking the clip, which will select it in Slip mode, hold the **Option** key and click on the clip to select it. You'll see the clip is selected differently from Slip mode; it's bracketed to the outside rather than the inside.
3. While holding the **Option** key, drag to the left to move the clip earlier in the Timeline (see Figure 6.17).

When you drag to the left, the clip in front of the clip being slid, *fresh tomatoes*, gets shorter and the clip after it, *prosciutto*, gets longer. When you slide to the right, the clip before gets longer and the clip after gets shorter. It's a bit like doing two Roll edits at the same time.

Once you have moused down on the clip with the **Option** key and started to drag it, you should release the **Option** key. When you do, the two-up display will appear in the Viewer, the shot prior, *fresh tomatoes*, on the left, and the shot after, *prosciutto*, on the right.

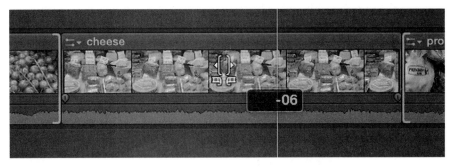

FIGURE 6.17

Sliding a Clip in the Timeline.

Just as in the other trim functions, you can use the nudge keys, **comma** or **period**, to nudge the clip up or down the Timeline. Make sure the clip is selected by clicking on it with the Trim tool while holding the **Option** key. The clip is then selected with the outward pointing brackets, and you can nudge or numerically move the clip as much as you like or as much as the available media of the two adjacent shots allows.

The Position Tool

One of the primary features of FCP is the Magnetic Timeline. All these edits have been controlled and affected to some degree or other by the Magnetic Timeline, particularly the Ripple tool. But sometimes you just want to shorten a clip and not move the adjacent clip. When you shortened a clip in earlier versions of the FCP and FCE Timeline, a gap would be created. To emulate this behavior, you have to use the Position tool, also known as the Pretend FCP7 tool. Let's shorten the *ms ham* shot:

1. To activate the Position tool, select it from the **Tools** menu in the Toolbar (see Figure 6.18) or press the **P** key.
2. Go to the end of the clip and you'll see a different cursor (see Figure 6.19), one that can affect only one side of the edit and will not ripple the sequence.
3. When you pull the end of the *ms ham* shot to the left, a hole will be created in the Timeline (see Figure 6.20). This is not a gap, as in previous versions of FCP, but is an actual object, a Gap Clip, that can be trimmed and replaced like any other clip in the Timeline. (Avid users will be familiar with this behavior.)
4. Click on the Gap Clip to select it and press the **Delete** key. The clip will be removed, and the Magnetic Timeline will close up the project.
5. Take the last clip, the *prosciutto samples* clip, and with the Position tool still active, drag it to the right, farther down the Timeline. A Gap Clip is created, separating it from the other clips in the project.

    These are all benign edits—shortening a clip, moving a clip to create a gap— but the Position tool has the potential to be very destructive as well.
6. Delete the Gap Clip you created in the Timeline.

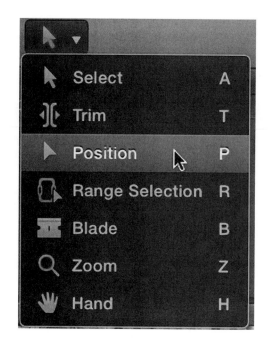

FIGURE 6.18

Position Tool.

FIGURE 6.19

Position Cursor in the Timeline.

FIGURE 6.20

Creating a Gap Clip in the Timeline.

7. Now drag *prosciutto samples* to the left, right on top of the previous clip. This is doing an Overwrite edit, wiping out part of the clip in the Timeline. (It's also effectively a Ripple edit of the previous shot, for which there are better tools with more control, as we have seen.)

8. Undo that edit with **Command-Z**.

   FCP has also built in similar destructive trim functions, which do not require the Position tool. Simply by selecting a clip and moving it numerically, you can wipe out other clips in the Timeline. You need to be very careful with this, as it's fairly easy to want to move the playhead and to mistakenly move a clip in a destructive manner because it's selected in the Timeline.

9. Return to the normal Select tool with the **A** key and click on the *ms ham* shot near the middle of the sequence to select it.

10. Type *+2.* (that's plus two and period), and the Dashboard will look like Figure 6.21.

11. Press **Enter** and the clip will move along the Timeline, wiping out two seconds of the next clip and leaving a two-second Gap Clip in the Timeline.

12. Undo the last edit with **Command-Z**.

You're not limited to moving a clip a few frames. You can use the move function to launch it anywhere in the Timeline, even down to the one-hour mark if you want to, leaving a Gap Clip in its place, and either cutting into other clips or creating more gaps farther down the Timeline. This is not a function you're going to need very often. In all the years I've been editing, I can't ever remember wanting to do this. Even if you don't want to do it, you should be aware of this behavior in the application, as it can be dangerous. You should also be aware that in addition to moving a clip numerically, you can also nudge a clip 1 frame or 10 frames with the comma and period keys, producing the same effect, moving the clip in the Timeline, wiping out part of the adjacent clip, and leaving a Gap Clip in the Timeline.

NOTE

Position Behavior

While the Position tool does allow a Roll edit between a clip and a Gap Clip, it does not allow you to do a Roll edit between two video clips.

FIGURE 6.21

Moving a Clip Using the Dashboard.

Trimming Connected Clips

Though most of the initial editing, at least of your A-roll, is done in the primary story-line, your B-roll is added as Connected Clips. Trimming these clips creates some issues in FCP because of the way the application handles these clips. For instance, if you have a string of Connected Clips above the primary storyline, you cannot ripple or roll the clips, nor can you slide the clips. You can, though, do a Slip edit using the Trim tool, as we have already done. Let's look at how to handle Connected Clips, but first we need to create some in our Timeline. A useful feature in some instances is the application's ability to lift clips from the primary storyline and make them Connected Clips, and also to take Connected Clips and overwrite them back into the storyline.

1. Select the last three clips in the project and use **Edit>Lift from Primary Storyline** or the keyboard shortcut **Option-Command-Up** arrow.

   The clips will move up and, if you select the middle Gap Clip, will look like Figure 6.22. Each clip is connected on its first frame to a separate Gap Clip underneath it. You cannot trim these clips as you normally would. If you drag a clip to shorten it, you will simply create a gap between two adjacent clips, as in Figure 6.23. If you try to push a clip to lengthen it, the adjacent clip will simply move out of the way onto a higher level, as in Figure 6.24. So how do we edit this material? You could simply drag clips to lengthen and shorten them and move them around as needed. Another, better way in most instances is to create an additional storyline.

2. Select all three Connected Clips and use **Clip>Create Storyline** or the useful shortcut that's well worth memorizing, **Command-G** (see Figure 6.25).

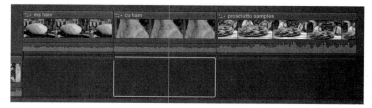

FIGURE 6.22

Clips Lifted from the Primary Storyline.

FIGURE 6.23

Gap Between Connected Clips.

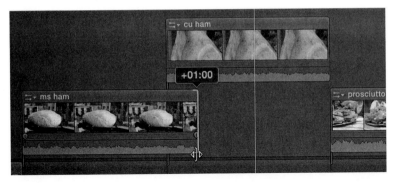

FIGURE 6.24

Connected Clip Pushed to Higher Level.

FIGURE 6.25

Additional Storyline.

TIP

Making Gap Clips Visible

Sometimes it's difficult to see Gap Clips, or it's clumsy to have multiple Gap Clips. Alex Gollner, who makes great FCP tools and effects (see www.alex4D.com), has a nice tip for this. Select the Gap Clip or clips and convert them to Compound Clips (**Option-G**). They will turn a pale gray and, while they're much easier to see and locate, still maintain the transparent properties of Gap Clips.

The clips are all joined together in a single storyline with the Shelf, the dark bar above the clips that serves to connect them. The storyline is now pinned to a single clip at the start. The other Gap Clips are redundant. The whole storyline can be picked up by the Shelf and dragged as a single group, as in Figure 6.26. Now any of the clips in the additional storyline can be trimmed and adjusted using any of the trim tools we have been using in the primary storyline.

Another technique that's very useful, especially when working with material like interviews or narration that's covered with B-roll, is to create an additional storyline immediately after the first Connected Clip is created. After the first Connected Clip is added and converted into a storyline, additional clips can be inserted, appended, or over-written directly to that storyline simply by selecting. So, for instance, in Figure 6.27, I have two storylines; if the primary storyline is my A-roll, I have an additional storyline that I have connected with a single point at the head of the Timeline.

FIGURE 6.26

Shifted Additional Storyline.

FIGURE 6.27

Additional Storyline with B-roll.

If I select the additional or secondary storyline, I can edit directly into it, keeping it a continuous additional storyline. If at any point I need to see the contents of the A-roll on the primary storyline, I can simply add a transparent Gap Clip into the additional storyline, making it however long I want. Any of the edits can be rippled and rolled, slipped and slid, on the additional storyline in relation to what's on the primary storyline. The additional storyline is my B-roll storyline and plays along with the A-roll, very much as the original telecine film chains did. This is a great technique for documentary, reality, and news programming, and can be used well for most productions that require B-roll to illustrate the story.

After you've adjusted your B-roll clips in the additional storyline, you can, if you wish, break apart the additional storyline at any time with **Clip > Break Apart Clip Items** or **Shift-Command-G**, the opposite shortcut from creating the storyline, as it were. The clips will then revert to being Connected Clips. Another option you have is to break apart the clips, not by breaking the storyline into separate clips, but by collapsing the additional storyline into the primary storyline. Simply select the storyline and use **Edit > Overwrite to Primary Storyline** or the shortcut **Option-Command-Down** arrow, the opposite shortcut from lifting from the primary storyline.

Additional storylines are a really useful and powerful feature in FCP and serve other uses, as we will see in the lesson on transitions.

THE PRECISION EDITOR

The Precision Editor will be familiar to iMovie users, but new to Final Cut users. The Precision Editor replaces some of the functions of the legacy Trim Edit window, though it does not have dynamic trimming capabilities.

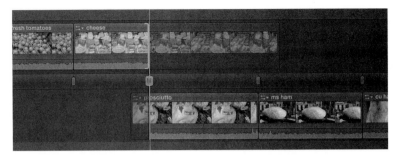

FIGURE 6.28

Precision Editor.

You can activate the Precision Editor in one of three ways:

- Select the edit and use **Clip>Show Precision Editor**.
- Select the edit and press **Control-E**.
- Simply double-click on an edit point.

Let's double-click on the edit point between the second and third shots, between *cheese* and *prosciutto*. The Timeline splits apart and the clips appear on separate tracks, the outgoing shot on the track above and the incoming shot on the track below with a dark separator bar between them (see Figure 6.28). The beauty of the Precision Editor is that you can see not only what's in the Timeline, but all the content of the clip, including the handles of extra media.

Notice the buttons on the separator bar. You can click on any of them to move to that edit point in the Precision Editor, or you can use the **semicolon** or **apostrophe** keys to move between edit points while keeping the Precision Editor open.

The pointer will change into different tools as you move around the window, and different clips will play depending on where the pointer is.

Let's do some editing in the Precision Editor:

1. Move the skimmer in front of the edit point on the upper track and press the **spacebar** to play. The outgoing clip will play through both the material in the Timeline and the available handles.
2. Move the pointer to the lower track and play that clip using the **JKL** keys to shuttle back and forth through the incoming clip.
3. To play through the edit itself, move the skimmer over the separator bar between the clips.
4. To roll the edit point, grab the button between the two clips on the separator bar and track it left or right.
5. To ripple either shot, you can drag the edge of the edit as you would in the Timeline, or you can simply grab the clip and push or pull it in the editor (see Figure 6.29).
6. You can also ripple either shot by moving the skimmer over either track and clicking on the clip. The playhead will move there, and the shot will be rippled to that point.

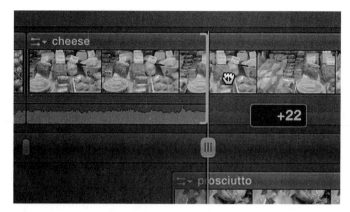

FIGURE 6.29

Dragging in the Precision Editor.

7. You can also use the Extend edit feature in the Precision Editor. This allows you to precisely position the skimmer using the play and arrow keys and then using **Shift-X** to execute the Extend edit. If a clip is selected on one side, a Ripple edit will be done. If both clips are selected in Roll mode, when you select the button between the two tracks, a Roll edit will be performed.

8. To close the Precision Editor, you can double-click on the separator button, press **Control-E**, use **Clip>Hide Precision Editor**, or perhaps simplest of all, just press the **Escape** key.

While the Precision Editor holds a great deal of potential, it doesn't offer quite the precision that most editors would like. Its reliance on the skimmer and its lack of dynamic trimming, where you can execute an edit while playing back, leaves something to be desired and to look forward to in further development.

TIP

Transitions

Transitions can be edited in the Precision Editor. The shots can be rippled; the edit point can be rolled; the transition can be made longer and shorter by dragging the handles. You cannot, however, open a transition edit point into the Precision Editor by double-clicking on it. You need to select it and use the keyboard shortcut **Control-E**.

NOTE

Clip Trimmer

There is no separate Clip Trimmer in FCP like the one iMovie users are familiar with. Those trimming functions are now built into the Timeline using the Trim tools we've already seen.

HOW LONG IS LONG ENOUGH?

A static shot, either a close-up or medium shot, must be on the screen a much shorter time than a long shot in which the audience is following a movement. A shot that has been seen before, a repeat, can be on the screen quite briefly for the audience to get the information. Though there is no hard-and-fast rule, because everything depends on the storytelling needs at the moment, generally, shots without dialogue remain on a television screen no more than six to eight seconds because of the small screen. In feature films, shots can be held for a lot longer because the viewer's eye has a lot more traveling to do to take in the full scope of the image. This is probably why older movies seem much slower on the television screen than they do in the theater. Although a close-up can be on the screen quite briefly, a long shot will often contain a great deal of information and needs to be held longer so viewers have time for their eyes to take it all in. You can often hold a moving shot such as a pan longer because the audience is basically looking at two shots: one at the beginning and the other at the end. If the movement is well shot—a fairly brisk move, no more than about five seconds—you can also cut it quite tightly. All you need to show is a brief glimpse of the static shot, the movement, and then cut out as soon as the camera settles at the end of the move.

SUMMARY

In this lesson we learned how to rearrange our shots by

- Using shuffle edits
- Reattaching clips
- Using Add edits and Gap Clips

We also saw how to trim our clips using

- Range Selection
- Trim Shortcuts

And we worked with the Trim Tools to do

- Ripple edits
- Roll edits
- Extend edits
- Slip edits
- Slide edits

We also used the Position tool and worked with Connected Clips and the Precision Editor.

Now you know the basic tools of editing with Final Cut Pro, cutting your clips, getting them into the Timeline, and trimming them. In the next lesson, we look at working with audio.

Working with Audio

7

LESSON OUTLINE

Film and video are primarily visual media. Oddly enough, though, the moment an edit occurs is often driven as much by the sound as by the picture. How many times have you heard a sound and turned to look in that direction? So let's look at sound editing in Final Cut Pro. How sound is used, where it comes in, and how long it lasts are key to good editing. With few exceptions, sound almost never cuts with the picture. Sometimes the sound comes first and then the picture, and sometimes the picture leads the sound. The principal reason video and audio are so often cut separately is that we see and hear quite differently. We see in cuts; for example, we look from one person to another, from one object to another, from the keyboard to the monitor. Though your head turns or your eyes travel across the room, you really see only the objects you're interested in looking at. We hear, on the other hand, in fades. You walk into a room, the door closes behind you, and the sound of the other room fades away. As a car approaches, the sound gets louder. Screams, gunshots, and doors slamming and other staccato, "spot" effects being exceptions, our aural perception is based on smooth transitions from one sound to another. Sounds, especially background sounds such as the ambient noise in a room, generally need to overlap to smooth out the jarring abruptness of a hard cut.

To overlap and layer sound we'll use Split Edits, which are incredibly easy to create in FCP, as the application flows the clips around each other. Nothing bumps against anything else; it simply moves out of the way, allowing us to easily cut video and audio separately from each other.

SETTING UP THE PROJECT

This instruction is going to sound familiar, but it's worth repeating. Begin by loading the material you need on the media hard drive of your computer. For this lesson we'll use the *FCP4.sparseimage*. If you haven't downloaded the material, or the project or event has become messed up, download it again from the www.fcpxbook.com website. Then do this:

1. Once the file is downloaded, copy or move it to the top level of your dedicated media drive.
2. Double-click the *FCP4.sparseimage* to mount the disk.
3. Launch the application.

The media files that accompany this book are heavily compressed H.264 files. Playing back this media requires a fast computer. If you have difficulty playing back the media, you should transcode it to proxy media:

1. To convert all the media, select the event in the Event Library, **Control**-click on it, and choose **Transcode Media**.
2. In the dialog that appears, check on **Create proxy media**.
3. Next, you have to switch to proxy playback. Go to **Preferences** in the **Final Cut Pro** menu.
4. In the **Playback** tab, click on **Use proxy media**. Be warned: If you have not created the proxy media, all the files will appear as offline, so you will have to switch back to **Use original or optimized media**.

The sparse disk has a couple of projects in its Project Library. We'll be working with these projects together with the event called *Audio* in the Event Library. Inside the event, you'll see the media we're going to work with.

Before we begin, I suggest duplicating the *Audio1* project:

1. Select the project on the *FCP4* disk in the Project Library and use **Command-D** to duplicate it.
2. Leave **Duplicate Project Only** checked and uncheck **Include Render Files**.
3. Duplicate the project to your system drive or your media drive outside the *FCP4* disk. This will preserve the original for you to return to if needed.
4. You can rename it to something like *Audio1 copy*.
5. Open the duplicate project.

THE SPLIT EDIT

A common method of editing is to first lay down the shots in scene order entirely as straight cuts, image, and audio. Look at the *Audio1 copy* project, which is the edited material cut as straight edits. What's most striking as you play it is how abruptly the audio changes at each shot. But audio and video seldom cut in parallel in a finished video, so you will have to offset them.

When audio and video have separate In and Out points that start at the same time, the edit is ~~~~~~~~ ~~~~~~~~ (see Figure ~~~~~), or an L-cut (see Figure 7.3 ~~~~~~~~~~~~~~~~~~~~~~~~~~~~~~~~~~~~~ them split edits.

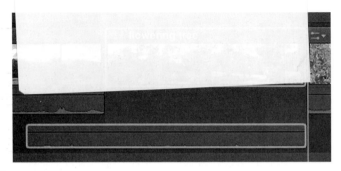

FIGURE 7.1

Split Edit.

FIGURE 7.2

J-Cut.

FIGURE 7.3

L-Cut.

The trick to smoothing out the audio for this type of project—or any project with abrupt sound changes at the edit points—is to overlap sounds and create sound beds that carry through other shots. Ideally, a wild track is shot on location; sometimes it is called *room tone* when it's the ambient sound indoors or *atmos* when it's outdoors. Atmos is a long section of continuous sound from the scene, 30 seconds or a minute or more, which can be used as a bed to which the sync sound is added as needed. Here, there is no wild track as such, but some of the shots are lengthy enough to have a similar effect.

Making Split Edits

Now let's make split edits:

1. Play through the first three or four shots in the project. The change in audio levels between the shots is quite noticeable.
2. To make it easier to manipulate the audio tracks, click the light switch in the lower right of the Timeline and select the second clip appearances option from the left (see Figure 7.4). Only choices 2, 3, and 4 are clip appearances that allow expansion.
3. Double-click the audio portion of the first clip *ls valley* to expand it. If it doesn't expand, you clicked in the upper part of the clip; in that case, move your cursor below the frame line and double-click again to expand audio. You can also use the keyboard shortcut **Control-S**, which will also collapse it if it's expanded.
4. Drag the tail of the audio edit point as far as it will go in the Timeline (see Figure 7.5) underneath the audio of the second clip. While you drag it, a small box will appear. It gives you a time duration change for the edit you are making.
5. There is a sharp spike of wind noise at the end of the clip, so drag the audio back a little to the left to clip that off.

FIGURE 7.4

Changing Timeline Clip Appearances.

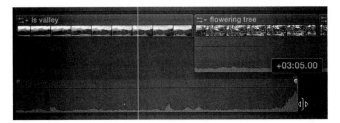

FIGURE 7.5

Creating Split Edit.

FIGURE 7.6

Audio Fade-Out.

6. In the upper-right corner of the audio track is a small fade button, which looks like an upside-down teardrop. With the pointer over the button, it will change to a drag tool. Drag it to the left to create a slow fadeout, as in Figure 7.6.
7. Double-click the audio portion of the second clip, *flowering tree*, to expand it, opening it below the first clip.
8. Drag the head of the audio to the right to stretch it out underneath *ls valley*.

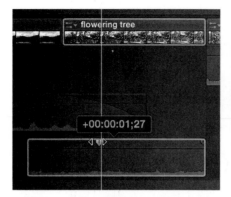

FIGURE 7.7

Overlapping Faded Audio.

FIGURE 7.8

Reducing the Level.

9. Drag the fade button at the head of the *flowering tree* audio to the left to fade that in, as in Figure 7.7.

10. Stretch out the end of the *flowering tree* audio as far as it will go and use the fade button to create a fade-out underneath *cs coffee flowers*.

11. The *drying beds* clip has poor audio with wind noise, and we don't need it at all. Grab the black level line that runs horizontally through the clip and drag it all the way down, as in Figure 7.8. We'll look at level changes in more detail in a moment.

12. One last step to clean up the project is to select the first two clips and press **Control-S** to collapse them. Though the fades are hidden, the sound transitions smoothly between the clips.

You have now created split edits in the Timeline, overlapped the audio, and cross-faded it. That's basically the process. Because the audio tracks always move out of the way for each other, it's really simple to overlap them and smooth out the audio transitions.

TRIMMING EXPANDED CLIPS

When clips are expanded, they can be trimmed independently. Normally, to select an edit point, as we saw in the preceding lesson, the **left** and **right bracket** keys and the **backslash** key will select the edit point, either in Ripple outgoing, Ripple incoming, or Roll mode. respectively. However when clips are expanded, as we've done here, these shortcuts will only select the video portion of the edit. Similarly, clicking on an edit will select only the video portion if you click the video and only the audio portion if you click the audio. If you want to select the audio only to ripple or roll the audio only via shortcuts, you can use **Shift-left bracket**, **Shift-right bracket**, and **Shift-backslash** to select the audio in, out, or center roll, respectively. You can then ripple or roll the edit as desired by nudging the edit, or numerically by typing a value, plus or minus, which will appear in the Dashboard.

> **TIP**
> ___
> Selecting an Audio-Only Edit
>
> For some reason you cannot select an audio-only edit in Roll mode. You can select it to ripple simply by clicking on it, but clicking with the Trim tool will not select it. Only the keyboard shortcut **Shift-backslash** will work.

Being able to roll the video independently of the audio is a very important feature, particularly when editing dialogue. The most common way to edit this material is to simply cut it in first as straight cuts showing whoever is speaking, getting the content into rough story order. In a documentary, this is sometimes called a "radio cut" or "a bed." In dramatic work, it's an "assembly edit." You see and hear, cut to cut, no B-roll, no overlaps. Once the best takes are in place showing the speaker or performers with the audio laid out and paced the way you want it, the video is then rolled separately so the picture and the sound overlap and you don't always see the person speaking when she speaks; rather you might hear the person before you see her or switch to someone listening before he responds, or cut to relevant B-roll. The possibilities are endless, and the best way you can to learn to edit, in addition to practice, is to watch the masters—well-shot and edited documentaries and narrative fiction—to see how scenes are expertly treated for image and audio experience.

CONTROLLING LEVELS

You can control the audio levels for clips in either the Event Browser or in the Timeline. It's easier and quicker and more common to control the levels after the clip is in the Timeline, but the Inspector controls for clips in the browser can be helpful, especially for long interviews that you're going to cut up and use several pieces. If you want to change the audio levels in the browser, you must change them before the clip is put into the Timeline. Changing the levels in the browser after the clip is in the project will not alter the levels of the Timeline clip.

Audio Meters

Whenever you work with audio, you need to see your audio levels. There are miniature audio meters in the Dashboard, which are really only there as confirmation that there is audio recorded. To see the real meters, either click on the tiny audio meters in the Dashboard or press **Shift-Command-8**. If your project is set to work with 5.1 surround sound, the meters will display six tracks for 5.1 sound monitoring. In standard stereo projects, the meters will show two tracks, as in Figure 7.9.

The standard audio level for digital audio is –12dB. Unlike analog audio, which has quite a bit of headroom and allows you to record sound above 0dB, in digital

FIGURE 7.9

Audio Meters.

recording, 0dB is an absolute. Sound cannot be recorded at a higher level because it gets clipped off. Very often on playback of very loud levels, the recording will seem to drop out completely and become inaudible as the levels are crushed beyond the range of digital audio's capabilities. It's important to keep sound at a good level: –12dB for normal speech, lower for soft or whispered sound, and higher for shouting. Normally, only very loud transients like gun shots will peak close to zero around –2dB or even –1dB. Make sure your audio does not bang up to the top of the meters or exceed 0dB so the LEDs at the top of the meters turn red, as in Figure 7.10. Now try this:

1. In the *Audio* event, select the clip called *rich* and play it back. The meters barely reach –20dB.
2. Open the Inspector by clicking the **i** button at the right edge of the Toolbar or by pressing the shortcut **Command-4**.
3. Switch to the Audio tab (see Figure 7.11). Here, you have controls for Volume and Pan, Audio Enhancements, and Channel Configuration.

TIP

Looping a Clip

To loop a clip, simply enable the Loop playback icon in the lower-right corner of the Viewer window, or press **Command-L**.

FIGURE 7.10

Overmodulated Audio Meters.

FIGURE 7.11

Audio Inspector.

Audio Enhancements

In the Audio controls, you could adjust the volume by pushing up the slider, but it would barely be enough even if pushed all the way up 12dB. Because the level is so weak in this instance, it's better to use the Audio Enhancements:

1. If the Audio Enhancements panel is hidden, double-click on it or click once on the **Show** button that will appear next to the hooked reset arrow if you move the pointer over it (see Figure 7.12).

FIGURE 7.12

Revealing Audio Enhancements.

FIGURE 7.13

Audio Enhancements.

2. Then click the button with the right-pointing arrow opposite Audio Analysis to open the Audio Enhancements panel. You can also open the Audio Enhancements panel by selecting **Audio Enhancements** from the Enhancements (the magic wand) pop-up menu in the Toolbar or by pressing the keyboard shortcut **Command-8**. The panel opens even if the Inspector is closed.

3. Check on the blue LED for **Loudness**, and you'll see that the audio is automatically increased 40% (see Figure 7.13).

4. If this still seems a little weak, pushing up the Loudness won't help much. Rather, increase the Uniformity to around 10%.

5. This setting might increase the background noise too much. Try checking on **Background Noise Removal**. I find the noise removal can be a bit too aggressive and adds *flanging* to the sound—that slight warbling effect as if the voice is underwater or in a tunnel or both.

6. Pull back the amount to about 30% or whatever suits your ear. The sound is not great, but you can get it to a level that's useable. The Loudness control in Audio Enhancements is very much analogous to the Normalization function in legacy Final Cut products.

7. Switch back to the main Audio tab with the arrow button in the upper-left corner of the panel.

8. Under Audio Enhancement, there is a line for Equalization with a pop-up next to it where you can select some standard equalizations, such as Voice Enhance, Music Enhance, Bass Reduce, and others.

9. Farther to the right, on the edge of the panel, is a small button icon with sliders. Click it to open the Equalization HUD (see Figure 7.14).

10. Using a button at the bottom, you can change the number of sliders from 10-band EQ to 31-band EQ. Adjusting the mids and the bass might help to enhance the voice.

There are other controls in the Audio Enhancement panel that are applied after an analysis:

1. In the browser, select the audio file *LatinWithHum*. If the Audio Enhancements panel is not open, press **Command-8**.

2. Notice that Hum Removal has been switched on. This happens automatically on import when analysis is switched on and AC hum is detected. Try switching off the Hum Removal, and you'll hear the low frequency 60-cycle hum that can easily be picked up from improperly grounded cabling.

3. Switch Hum Removal back on when you're done.

At the bottom of the Audio Enhancements panel, notice the button for Auto Enhance. This will check your audio and apply adjustments in the Audio panel that

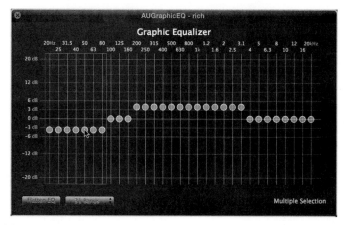

FIGURE 7.14

A 31-band EQ HUD.

FCP thinks will work for your clip. You can also do this using the Enhancements pop-up menu in the Toolbar and selecting Auto Audio Enhancements (see Figure 7.15) or using the keyboard shortcut **Option-Command-A**.

Channel Configurations

Unlike legacy versions of Final Cut Pro and Express where channel configurations, stereo pairs, and dual mono could be changed only after clips were in the Timeline, in FCP, Channel Configurations can be set while clips are still in the browser or when they're in the Timeline. Here's how:

1. Select the *measuring moisture* clip in the Event Browser.
2. Go to the Audio panel of the Inspector. The last item is Channel Configurations.
3. If you twirl the Channels open, you'll see the audio tracks displayed.
4. There is a pop-up that lets you set the channel configurations to either Stereo or Dual Mono or other configurations (see Figure 7.16).
5. There are also checkboxes that allow you to selectively switch off tracks as needed.

Channel Configurations can be changed not only to single clips, but to multiple selected clips in the Event Browser or in the Timeline.

FIGURE 7.15

Auto Audio Enhancements.

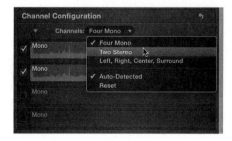

FIGURE 7.16

Channel Configurations.

> **TIP**
>
> Single-Track Audio
>
> Often you record audio on a single audio channel. Many cameras do this when you plug in an external microphone; only one of the stereo channels gets used. It's worth deleting the empty track. FCP will detect empty audio tracks on import and remove them automatically. However, sometimes the second track isn't actually empty, just poor quality from a distant camera mic. If the audio is a stereo pair, first change it to dual mono. Unlike in earlier versions of FCP, in this version you can do this before the material is edited into the project. Also unlike in legacy Final Cut, in FCP the material will automatically be centered, although obviously you can alter the pan values as you like.

Mixing Levels in the Timeline

Let's look at adjusting levels and mixing audio in the FCP Timeline panel. There is no separate audio mixer as such, but you have direct control of the audio levels of your media in the project. We'll work in the *Audio2* project and start by duplicating it.

1. Select the project on the *FCP4* disk in the Project Library and use **Command-D** to duplicate it.
2. Leave Duplicate Project Only checked and uncheck Include Render Files.
3. Duplicate the project to your system drive or your media drive outside the *FCP4* disk. This will preserve the original for you to return to if needed.
4. You can rename it to something like *Audio2 copy*.
5. Open the duplicate project.

Play through the last few clips in the sequence to listen to it and hear some of the problems. We already know that Rich's audio is too low; we've already fixed that in the browser. You don't have to do that again; you can use the same settings you used in the browser for the clip in the Timeline. The audio all needs to be smoothed out, as we did earlier by overlapping the clips, and the last couple of shots are too loud. We also want to add some music at the beginning and a couple of more shots to open it up a bit. Let's add the music to the beginning first. I like to have everything in the Timeline if I can before I start mixing the sound. We could attach the music to the first clip, but we want to add more shots; plus, I always think it's a good idea to begin the music over black and attach the music to that.

1. Put the playhead at the beginning of the Timeline and use **Edit>Insert Gap** or the handy shortcut **Option-W**. The insert gap is three seconds, but you can make it any length you want.
2. With the playhead still at the beginning of the Timeline, find the *LatinWithHum* clip in the *Audio* event and press the **Q** key to attach it to the Gap Clip at the beginning of the project. The playhead will zoom to the end of the Timeline, and everything will disappear because the music is much longer than the picture.
3. Press the **Home** key (or **fn-Left** arrow on a laptop) to go back to the beginning of the project.

4. In the browser, find the *cottage* shot and mark a selection beginning just after the start of the clip, leaving the duration about two seconds.
5. Move the pointer to the end of the Gap Clip so it changes it to the Ripple function and push the gap so it's about five seconds long.
6. Play the Timeline for the first few beats to the beat at 1:06. We're going to leave the first bit of black in the Timeline, but we're going to overwrite the *cottage* clip into the gap.
7. Press **Option-2** so that you do a video-only edit. We don't want to hear the voices talking on the *cottage* shot.
8. Press the **D** key to overwrite the shot into the primary storyline.
9. Select the Gap Clip that remains after *cottage* and before *rich* and delete it.
10. With the video-only edit still on, insert a couple of seconds of the *dalmatian* shot after the *cottage* with the **W** key.

Fading Levels in the Timeline

We're about ready to start working on our audio. So you can readily create split ends and the overlaps you want, often the easiest way to do this is just to expand everything in the Timeline, but before we do that, we'll add in one more piece of sound:

1. In the Timeline, move the playhead so you're one second into the project, still in the black of the Gap Clip, and press **I** to mark the beginning of a selection.
2. Play forward till you're a second or two past the beginning of the *ls valley* shot and press **O** to mark the end of the selection.
3. Press **Option-3** to prepare to make an audio-only edit.
4. Find the *ls valley* shot in the browser, select it, and press **Q** to connect the audio to Gap Clip underneath the music.
5. To expand all the Timeline clips, select all the clips in the Timeline with **Command-A** and then press **Control-S** to expand all the clips. All the clips that have audio will now be split apart.
6. Grab the fade button at the beginning of the single *ls valley* audio clip and drag it to the right to get a nice fade-up of the birds singing. The farther you drag the fade, the slower it will be.
7. **Control**-click on the fade button, and a HUD will appear, allowing you to select the type of fade you want to use (see Figure 7.17). Select the Linear fade.

The default curved or logarithmic ramp is the +3dB fade, which is perfect if you are cross-fading between overlapping sounds as we did in the first project. This gives what's called an equal power fade where there is no apparent dip in the audio level in the middle of the crossing fades. Generally, use a Linear 0dB fade when you are fading up or fading down to silence, as we are here. There are four fade types to choose from, and you can use whatever's right for the sound you're working with. Sometimes fading in or out of speech with an S fade or a −3dB fade might work better.

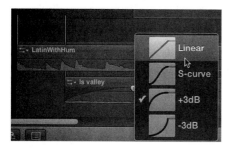

FIGURE 7.17

Fade HUD.

FIGURE 7.18

Solo Button.

Before you mix the body of the project, you should set the level for the primary audio, which in this case is on the primary storyline. But we've already fixed the audio for this clip once in the browser, so we don't have to do it again.

1. To make it easier to hear the clip, select *rich* in the Timeline. Then click the Solo button in the upper right of the Timeline so it becomes yellow (see Figure 7.18); or press **Option-S**, use **Clip>Solo**, or right-click over the clip and choose the **Solo** option.
2. Now find the *rich* clip in the Event Browser with **Shift-F** (Reveal in Event Browser), and select **Copy** from the **Edit** menu, or use **Command-C**.
3. Click back on *rich* in the Timeline and use **Edit>Paste Effects** or the shortcut **Option-Command-V** to paste the same audio attributes applied to the browser clip to the clip in the project.
4. Check the audio and then unsolo the clip with **Option-S**.

> **NOTE**
>
> Marked Selection
>
> In legacy versions of Final Cut when the match frame function was evoked, the equivalent of Reveal in Event Browser, the same selection that was on the clip in the Timeline would appear as a selection on the browser clip. FCP does this as well, but only if the browser is set to Filmstrip view. In Filmstrip mode, a yellow, marked selection area will appear that matches the clip in the Timeline. However, if the browser is in List view, the selection will not appear; only the whole clip will appear in the preview at the top of the list.

Changing Levels on Part of a Clip

There may be occasions when you want to change the audio level of a section of a piece of audio, reduce a section of music while someone is speaking, or raise a section where the level is too low. We want to reduce, or "duck," the level of the music as Rich starts to speak and then, after a while, fade it out completely. This approach often works better than a simple, straight fade-out. To select the area to reduce, we're going to use the Range Selection tool. This should be a technique familiar to iMovie users, but something a little novel for Final Cut users.

1. Activate the Range Selection from the Tools menu in the Toolbar or by pressing the **R** key (see Figure 7.19).
2. Drag an area on the *LatinWithHum* clip in the Timeline, beginning shortly before the voice begins until about 11 seconds into the project.
3. With that area selected, pull down the level line so the audio is reduced about −11 or −12dB. Keyframes will automatically be added, ramping the sound down and back up at the end of range selection (see Figure 7.20).
4. Return to the Select tool by pressing the **A** key, and either click off the clip to drop the selection or use **Shift-Command-A**.
5. To slow down the fade at the start, pull the outside keyframes farther apart.

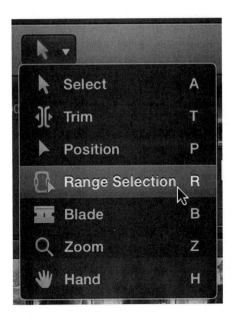

FIGURE 7.19

Range Selection Tool.

FIGURE 7.20

Audio with Keyframes.

6. At the end of the reduced audio section, drag the final keyframe farther along the Timeline and then drag it down so the audio fades out completely.

7. With the Blade tool, cut off the extra length of the music that you don't need.

NOTE

What Is a Keyframe?

We'll be talking more about keyframes as we get further into the book. A keyframe is a way of defining the values for a clip at a specific moment in time, a specific frame of video. Here, we're dealing with audio levels, so we're saying that at this frame, we want the sound to be at a particular level. By then going to a different point in the clip and altering the levels, we will have created another keyframe that defines the sound level at that moment. The computer will figure out how quickly it needs to change the levels to get from one setting to the other. The closer together the keyframes are, the more quickly the levels will change; the farther apart they are, the more gradually the change will take place.

If you need to add additional keyframes to raise or lower the audio levels, simply move the pointer to the level line, hold the **Option** key, and click on the level line. A new keyframe will be added. You can add as many as you need, raising and lowering the audio as required. If you want to delete a keyframe, simply click on it to select it and press the **Delete** key.

TIP

Audio Views

When you're working with audio, it's sometimes easier to work with just the audio waveforms and not bother with video. You can do this in the Clip Appearance switch in the lower right of the Timeline. The first selection gives you large tracks and large waveforms and might be best for the work we're doing. Even though only the audio tracks and their waveforms appear in the Timeline, the video still appears and plays in the Viewer when the tracks are skimmed.

Setting Clip Levels

Before creating the overlaps in the tracks to cross-fade the audio, let's set the levels for the Connected Clips. If you click on each clip in turn, you'll notice in the Audio tab that some indicate they need some enhancement; they need background noise reduction. You don't have to do each one.

1. Simply drag-select to highlight all the Connected Clips, and from the Enhancements pop-up in the Toolbar, select **Auto Enhance Audio** or use **Option-Command-A**. Done. Noise reduction or other enhancements are applied to all the clips that need it.

2. With the clips still selected, in the Audio tab, drag the Volume slider to the left to reduce the overall level of all the clips. This will bring down the background sound.

3. Adjust the levels on the two last clips in the project that have the loud coffee roasting sound.

A great way to adjust the levels of a clip is to do it dynamically while it's playing back. If you select a clip in the Timeline (use the **X** key to select each clip as the cursor passes over), you can raise and lower the audio levels with the keyboard shortcuts **Control-** (minus) and **Control-** = (think of it as plus). The first will lower the level by 1dB, and the second will raise it 1dB. The great thing about these shortcuts is the audio can be adjusted during playback. This capability is especially useful for setting the level for music or an interview.

NOTE

Connected Audio in the Timeline

Because there are no tracks in FCP as understood in legacy versions and other applications, audio can be put anywhere in the Timeline. When you connect a piece of audio to the primary storyline, it is connected underneath the storyline, but this is really for no more reason than it's traditional. If you drag an audio clip into the project, you can put it above the video if you want. This approach does make for a messy Timeline, but it works perfectly well.

TIP

Reference Waveform

When you're reducing the audio levels of clips in the Timeline, especially to low levels, it's handy to still see what the waveform looks like even when the audio is very low. You can do this by checking on **Show reference waveforms** next to Audio in the Editing tab or user **Preferences**. With reference waveforms switched on, you see a ghost of the waveform in the Timeline, as in Figure 7.21.

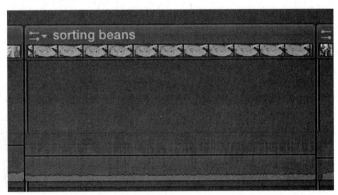

FIGURE 7.21

Reference Waveforms.

Match Audio

One of the great new features of FCP is the Match Audio function. This compares the audio equalization of one clip and balances it to another by applying EQ settings. Users of Soundtrack Pro will be familiar with this function. Here's how to use it:

1. In the Timeline, select the *luna sign* shot, the third from the last Connected Clip.
2. From the Enhancements pop-up menu in the Toolbar, select **Match Audio** or use **Shift-Command-M**. A two-up display appears in the Viewer (see Figure 7.22).
3. Move the skimmer over the shot before *luna sign*, the *cs coffee flowers* shot. The pointer will change to an EQ icon. Click on the *cs coffee flowers* shot.
4. In the Viewer, click the **Apply Match** button.

If you want to change it, you can click the **Choose** button that will appear opposite Audio Enhancements in the Audio tab. If you want to see what equalization has been applied to the clip, click the EQ icon next to it to bring up the HUD that controls the effects (see Figure 7.23). You can make any adjustments you want in the HUD to fine-tune the audio.

You can even animate the equalization using keyframes if you want. Every clip in the Timeline has a tiny badge in its upper-left corner that is actually a

FIGURE 7.22

Match Audio Display.

FIGURE 7.23

EQ HUD.

FIGURE 7.24

Timeline Clip Badge.

pop-up that gives you access to animation controls (see Figure 7.24). Here's how to use it:

1. Click the animation badge for *luna sign* and select **Show Audio Animation** or use the shortcut **Control-A**.
2. If you click on Match EQ that appears in the green animation bar below the clip, you will see a level line that you can **Option**-click to set keyframes.
3. The disclosure triangle will give you a list of properties you can animate, from All to very specific parameters (see Figure 7.25).
4. Close the animation display by using the little **X** button in the upper left of the clip.

Subframe Precision

One of the features for editing sound in FCP is that you can do it with great precision, down to subframe accurate detail. To help you do this, you can switch the Dashboard

FIGURE 7.25

Audio Animation Controls.

FIGURE 7.26

Subframe Preference.

FIGURE 7.27

Subframe Audio Edit.

display to show subframe amounts. In user **Preferences**, go to the **Editing** tab, and from the **Time Display** pop-up, select **HH:MM:SS:FF+Subframes** (see Figure 7.26).

With subframes displayed in the Dashboard, you can use the **Command** key and the **Left** or **Right** arrows to move the playhead in increments of 1/80th of a frame.

If you must zoom in to the waveform in the Timeline, you can set keyframes within a single frame of video. This capability is really useful if there are spikes on the audio you need to reduce or single sounds you want to edit out. In Figure 7.27 the broad pale area across the image is a single frame of video. As you can see, I could cut down a very small section of that 1/30th of a second to trim out a narrow sliver of audio.

Once you've finished mixing your audio and setting the levels, it's always a good idea to select the clips and collapse them again. You can do this from the **Clip** menu or with the same shortcut used to expand the clips, **Control-S**.

TIP

Zoom a Selection

You can use the Zoom tool (**Z**) to drag a selection around a section of the waveform to zoom in to just that portion of the display.

PAN LEVELS

In the Inspector, in addition to controlling the volume of a clip, you can also control the Pan mode, which defaults to None. With the pop-up, you can change it to Stereo Left/Right or other settings. When Stereo Left/Right is set, as in Figure 7.28, you have a Pan Amount slider that allows you to move the audio from left to right as you like and to animate it to move from the left or right speaker across the screen. For fun, let's do this to the music in the Timeline. You can do this entirely in the Timeline, but we'll start off by setting the first keyframe in the Inspector:

1. Select *LatinWithHum* in the Timeline. Then move the playhead right to the beginning of the project with the **Home** key and go to the Inspector panel.
2. With Stereo Left/Right selected, move the Pan Amount slider all the way to the left to –100 and click the **Add Keyframe** button on the right end to the slider.
3. Move the Timeline playhead to 2:15 and click the **Add Keyframe** button again. It's important that you click the **Add Keyframe** button first, or you will simply be changing the value of the first keyframe. Notice little arrows to one or both sides of the keyframe to navigate between them.
4. Move the Pan Amount slider all the way to the right to 100, and listen to the music. You'll hear it move from the left speaker to the right speaker. Notice the audio line is now black rather than white during the pan change.
5. Click the Animation badge for the clip in the Timeline or press **Control-A** for Audio Animations.
6. Double-click Pan Amount in the Timeline to open the keyframe graph for that parameter (see Figure 7.29).
7. Move the playhead forward to about 5:00, and holding the **Option** key, click on the Pan Amount level line to add another keyframe.
8. Drag the third keyframe back down to zero. The music will move from left to right and then come back to be center-panned.

Here, the effect is pretty silly, but it's often useful to move sound from one side to the other—for instance, for a car that drives across the screen.

In the Pan mode pop-up, there are also presets for 5.1 surround panning, such as Dialogue, Music, or Ambience, which spreads the sound to the periphery and leaves the center open for voice, as in Figure 7.30 (also in the color section). Notice the puck in the Surround Panner that can be revealed with the disclosure triangle. By dragging the puck around the Surround Panner, you can control the position of the audio in the surround space. Not only is the position of the puck keyframeable, but so are all

FIGURE 7.28

Stereo Left/Right Pan.

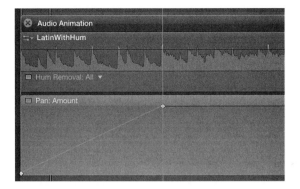

FIGURE 7.29

Pan Keyframe Graph in the Timeline.

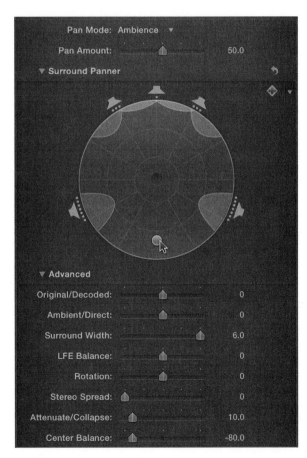

FIGURE 7.30

5.1 Surround Panner and Controls.

the advanced properties such as Surround Width, LFE Balance, and Rotation. With a proper surround monitoring system, you can lay out multiple tracks and assign an amazing variety of surround options to control your audio.

ROLES

Because FCP has no tracks, and objects on different layers move around to accommodate other objects, keeping the content organized becomes very difficult. It's important so that you can export *stems* from your project. Stems are separate tracks of audio that are used to mix together different types of sound for different delivery purposes. So you might have separate stems for your sound effects tracks, for your music tracks, for your dialogue, and perhaps even separate stems for each character's dialogue so that the audio can be mixed and equalized separately when it's sent to a professional audio mixing suite. Obviously, this is a problem in FCP because of the way it works. To make the creation of stems possible, FCP uses **roles**, which are metatags assigned to clips specifying what stem that clip belongs to.

You can set the role for a clip in the Event Browser before the clip is edited into the Timeline, or you can set it in the Timeline after the clip has been edited into the project. Often roles aren't assigned until after the clips are in the project; this is especially true if you're going to use custom roles and subroles, which are subsets of roles.

> **NOTE**
>
> Project Clips Are Separate
>
> If you set a clips role in the browser and then edit it into the Timeline, the role will go with it. If you assign the role in the Timeline, the browser clip is not changed. Also if you change the role of a clip in the browser, the copy of the clip that's in the Timeline does not have its role changed.

There are two types of video roles: one is Video, and the other is Titles. There are three basic types of audio roles: Dialogue, Music, and Effects. By default, any video clip imported has two roles assigned to it: Video and Dialogue. Any music that's imported from iTunes has Music set as its role. Obviously, you're going to want to change these. Fortunately, you don't have to change each one separately, but you can select groups of clips and change them once. So, for instance, you could select all your B-roll video that's in keyword collections and change their roles to Video and Effects. Let's change the roles for our B-roll in the Timeline.

1. Select all the B-roll Connected Clips by dragging through them.
2. With the clips selected, go to the Info tab of the Inspector, and from the Roles pop-up, select Effects (see Figure 7.31). Notice the easy keyboard shortcuts: **Control-Option** plus the first letter of the role you want to select.
3. Select the audio effect clip in the project and change that to **Effects**, and change the music in the Timeline to **Music**.

FIGURE 7.31

Select Roles.

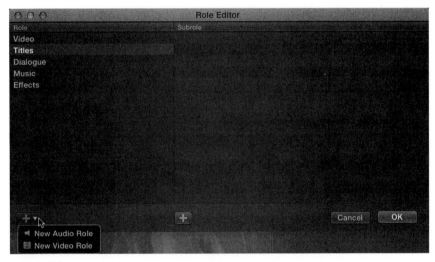

FIGURE 7.32

Editing Roles.

The two first video clips have only Video selected as their role, and Rich has Video and Dialogue as his roles, which are correct. That's it.

In the Roles pop-up in the Info tab, notice the option **Edit Roles**. This brings up the window in Figure 7.32. There are two + buttons at the bottom. The one on the left allows you to add roles, and the one in the middle lets you create subroles of a selected role. Subroles will let you add separate roles for each character or type of music or type of effect.

EDITING MUSIC

Editing music is very different in Final Cut Pro from earlier applications, and will not be very familiar except in concept to either Final Cut or iMovie users. One thing iMovie users will be familiar with is the application's ability to access audio

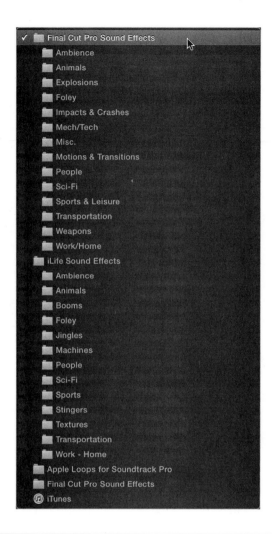

FIGURE 7.33

Audio Browser Content.

content using the media browser. Clicking on the music note icon in the media browser on the right side of the Toolbar will give you the Final Cut Pro and iLife sound effects, loops, and jingles (see Figure 7.33). Also notice you can access any of your iTunes content. At the bottom of the audio browser (see Figure 7.34) is a search box with which you can search the content. This capability is especially useful when looking for sound effects. The play button next to the search box lets you listen to the audio.

When you've found what you want, simply drag it into the project or into an event or even a Keyword Collection. If you drag the music or effects into a project, it is also automatically added to the default event for that project.

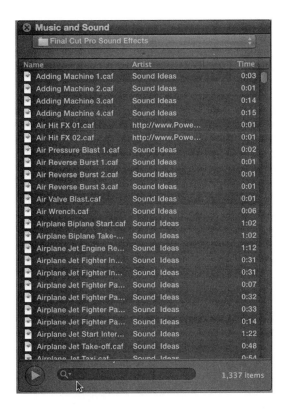

FIGURE 7.34

Music and Sound Effects Search.

Adding Markers

One of the best ways to edit music is to use markers. You can add markers to a clip either when it's in the Event Browser before you put it in the Timeline or to a clip after you've put it into the Timeline. Just as in iMovie and earlier versions of Final Cut, you can add markers to the beat by tapping the **M** key. You should be aware that because audio-only files like *LatinWithHum* have no real timebase in the browser, they are treated as 59.94 frames per second material (60fps). In the Timeline, however, they use the timebase of the project they're in. So let's make a new project and edit some music:

1. Go to the Project Library by clicking the film reel in the lower left of the Timeline or pressing **Command-zero**.
2. Make sure your system or your media drive is selected, not the *FCP4* disk image, and press **Command-N** to make a new project.
3. Name the project *Music*, set *Audio* as the default event, and set Audio Properties to **Custom, Stereo**.
4. Select *LatinWithHum* in the *Audio* Event Browser and press **E** to append it into the Timeline.

5. Because this is an audio-only clip, a dialog will appear, asking you to set the project properties. Make the Video **1080i HD**, set the Resolution to **1920×1080**, and set the Rate to **29.97i**. You use these settings because they match the video we're working with.

The music will appear in the primary storyline ready to have clips connected to it. We don't have enough video for all the music, but we'll put a few shots into the project. But first let's add the markers:

1. Move the playhead back to the beginning of the sequence with the **Home** key and zoom in with **Command-plus** so it's easier to see the waveform.

2. You can either play the project and tap the **M** key as you play, or you can add the markers at specific frames. Put the first marker at the beat at 1:07 in the Timeline.

3. Add the next marker at the beat at 2:14 and zoom in so you can clearly see one frame.

4. Holding the **Control** key, use the **comma** and **period** keys to nudge the marker left and right on the clip in 1/80th of a frame increments (see Figure 7.35). In reality, this precision is not going to help you much with editing your video, as video can be edited only in full-frame increments.

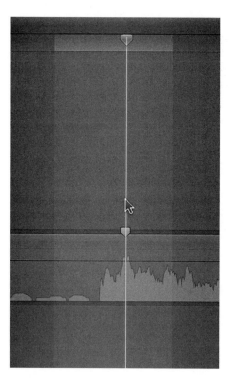

FIGURE 7.35

Nudging Markers.

5. If the marker is in the wrong place, you can delete it by going to the marker and pressing **Control-M**. You can delete all the markers on a selected clip or a selected range of a clip with **Control-Shift-M**.

6. You can name markers in the marker HUD by either pressing the **M** key when you're on a marker or simply double-clicking it (see Figure 7.36).

7. Go down to the beat at 3:20 and press **Option-M**. This not only adds a marker, but also opens the marker HUD. A double-tap on **M** will do the same thing.

8. Click the **Make To Do Item** button to give yourself a reminder. The marker will turn red.

9. To go to a previous marker, use **Control-semicolon,** and to go to the next marker, use **Control-apostrophe**.

10. Add more markers at 4:27, 6:04, 7:10, 8:17, 9:15, and 10:08.

Table 7.1 gives a list of useful keyboard shortcuts for working with markers.

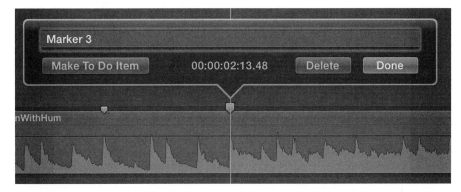

FIGURE 7.36

Marker HUD.

Table 7.1 Marker Shortcuts

Marker Tools	Shortcut
Add marker	M
Add marker and open dialog	Option-M
Open Marker dialog	Shift-M or M
Delete marker	Control-M
Delete markers in selection	Control-Shift-M
Nudge a marker left	Control-comma
Nudge a marker right	Control-period
Go to previous marker	Control-semicolon
Go to next marker	Control-apostrophe

> **NOTE**
>
> Chapter Markers
>
> Though you can add markers to clips in FCP, you cannot add markers to the Timeline Ruler as you used to be able to do in FCE or earlier versions of FCP. Because of this, you cannot create chapter markers for DVDs or for web video in FCP. To add chapter markers, you have to use Compressor, or some people like to use iMovie. Be careful of how you work with iMovie, especially with standard definition media, as your video deteriorates substantially.

CONNECTING CLIPS

Now we're ready to add some clips to the project. Using the markers as timing points, we can connect clips to the music on the primary storyline. Normally, I would cut audio and video into the project, but in this case I think we'll edit picture only. Let's start by marking a selection in the Timeline:

1. Move the playhead to the first marker. The easiest way is with the Timeline active to press the **Home** key and then **Control-apostrophe**.
2. Mark the end of a selection with the **O** key. The selection will be made from the beginning of the clip to the marker, but the selection is one frame too long.
3. Because the selection includes the frame you're on, you either have to step back one frame with the **Left** arrow key and press **O** again, or simply click in the selection range and the selection will be pulled in one frame with the Range Selection tool.
4. Press **Option-2** or select **Video Only** from the edit button pop-up in the Toolbar to do an edit without adding the sound to the project.
5. In the *Audio* Event Browser, find the *coffee cups* shot and mark a selection start point with the **I** key a couple of seconds into the shot.
6. Press **Q** or the Connect edit button to connect the clip and add it to the Timeline (see Figure 7.37).
7. You're ready to repeat the procedure. In the Timeline (**Command-2**), mark an In point at the first marker, move to the second marker, go back one frame, and add an Out point.

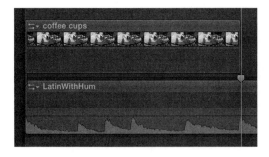

FIGURE 7.37

Connected Clip at Marker.

8. In the browser, mark an In point early in the *cooling roast* shot and press **Q** to connect it.

9. Add an In at the second Timeline marker and an Out one frame before the third marker.

10. Find *luna sign* in the browser and mark an Out point just before the end, just before the camera swings down to the floor.

11. Press **Shift-Q** to backtime the shot into the Timeline from the end of the Timeline selection.

12. Continue to the other marker segments at *dalmatian, flowering tree, cottage, sorting beans, drying beds*, and *rich and roaster*, or any other shots in any other order that you want.

You could also start to edit your music by putting a Gap Clip into the Timeline and connecting the music to that and then editing your video into the primary storyline. I prefer the method I described, but you can use whichever works for you.

TIMELINE INDEX

Though this isn't strictly part of editing music, it seems like a good opportunity to show you the uses of the Timeline Index. It can be used to search the content of your project as well as to navigate in it.

Let's switch back to the *Audio2 copy* project we were working on, which has defined roles for the clips.

TIP

Switching Between Projects

If you have been working with multiple projects, clicking and holding one of the history arrows in the upper left of the Timeline will give you a pop-up list of recently opened Projects.

FIGURE 7.38

Roles Index.

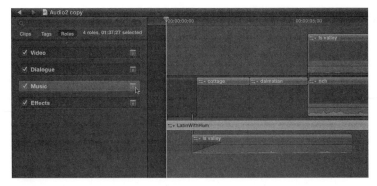

FIGURE 7.39

Minimize Roles.

FIGURE 7.40

Disabled Roles.

To open the Timeline Index, either click the little index button in the lower left of the Timeline panel or press the shortcut **Shift-Command-2**.

The index has three tabs at the top for Clips, Tags, and Roles. The index opens on the Roles tab, which allows you to control the Timeline display. Selecting a role will highlight it in the Timeline (see Figure 7.38, also in the color section). Clicking the button on the end (see Figure 7.39) will minimize the role, so it's dark in the Timeline. Clicking the checkbox at the head of the role (see Figure 7.40) will switch off the role completely. Here, the project will only play back and export the video and the dialogue roles; music and effects have been switched off.

Make sure all the roles are switched on and click on the Clips tab of the Timeline Index. Here, the index shows the list of clips used together with their timecode locations (see Figure 7.41). To move the playhead to a clip in the project, simply click on its name. Notice there is a search box at the top, and four buttons at the bottom that let you filter to display just video clips, audio clips, or just titles.

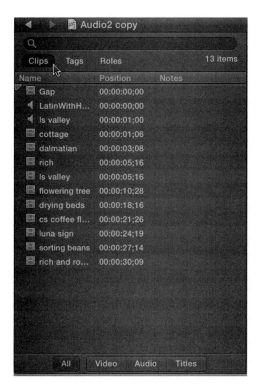

FIGURE 7.41

Timeline Index.

TIP

Gap Clips

Gap Clips appear in the Timeline Index as *Gap*. To perhaps make them more helpful, you can select a Gap Clip in the Timeline and use the Info tab of the Inspector to change its name to something relevant that will be meaningful in the Timeline Index.

Tags are markers. To see the Tags tab properly, switch back to the *Music* project where we added markers. You should be able to do this with the back button in the upper left of the Timeline. Click the Tags button to switch to the Tags index (see Figure 7.42). Here again, you have a search box at the top, as well as buttons at the bottom that let you see all tags, markers, keywords, analysis keywords, To Do items, and completed items.

1. Click the To Do tags, the second button from the right, and you should see the one To Do item we made.
2. Click on the red button next to it, and it will disappear from the To Do list. It will be in the Completed list, and the marker in the Timeline will go green.

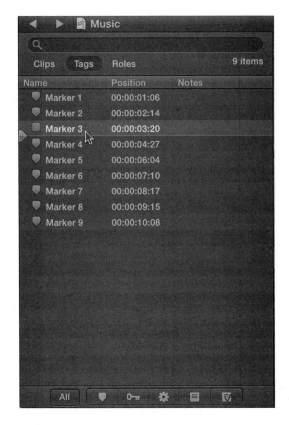

FIGURE 7.42

Tags Index.

You can also set To Do items to be completed in the marker HUD. The Timeline Index becomes more and more useful the longer your project gets. It's a very handy way to navigate through your material and to keep track of things you have to do.

COMPOUND CLIPS AND MIXING

Most advanced video editing applications have some audio mixing capabilities aside from simply keyframing individual clips. To facilitate mixing like sounds such as music, primary narration, or interviews, many applications have track grouping and busing to apply effects to multiple items simultaneously. Once a picture has been locked, you can group similar items together, selecting them and using **File>New Compound Clip** or **Option-G**. The Compound Clip can be renamed in the Info tab of the Inspector and can be treated as a separate clip. The audio levels can be adjusted globally, and audio effects such as a Limiter can be added and adjusted. Should you need to control the clips inside the Compound Clip separately, you can simply double-click it to open it into the Timeline window. Anything you do inside the Compound Clip you created in a project will be reflected in the final project. Compound Clips share some of the features of nesting in legacy FCP.

SYNCHRONIZING CLIPS

More and more people are shooting video with DSLR cameras. One feature of these cameras is their very poor audio recording capabilities, compounded by the complete inability to monitor audio, which are two fundamental functions of any video camera. Because of this, many users are recording audio using a separate recorder like a Zoom recorder. FCP now has the ability to automatically sync up picture and sound by looking at the audio waveforms of the two items. There are two items in the *Audio* browser that aren't named; they simply have dates. One is a video file that's in the ProRes Proxy codec, and the other is an audio file that corresponds to it.

1. Play the video file. That's me, by the way. There's a handclap at the beginning for syncing, and if you're monitoring the meters, you'll see they barely rise above –30dB. Raising the level and turning on Loudness will make the level acceptable but will also bring up a lot of hiss.
2. Play the audio file. That's much better; in fact, it probably should be reduced as it's a little too loud.
3. Select both items in the browser and use **Clip>Synchronize Clips** or **Option-Command-G**.
4. A new item called *Synchronized Clip: 2011-09-06_12_35_38* is created. That's a bit unwieldy, so rename it *SyncTom*.
5. Either double-click the sync clip to open it into the Timeline or **Control**-click on it and select **Open in Timeline**. Here, you can see how the video and audio line up, and slide them independently to adjust the sync if necessary (see Figure 7.43).
6. Select the video portion. Then, in the Audio tab of the Inspector, go to Channel Configuration and uncheck the camera stereo sound to switch it off.
7. You can trim the beginning and end of the sync clip, and this will be reflected in the browser. If you do trim it, the end of the sync clip will go into black, so you'll need to be careful when editing it.

You are now ready to edit your sync clip into your project. To return to the project, click the **Back** button, the triangle in the upper-left corner of the Timeline panel. There are buttons here for forward and backward, which allow you to easily switch between recently opened projects, clips, or compound clips, navigating through the

FIGURE 7.43

Clip Opened in Timeline.

Timeline History. You can use the same keyboard shortcuts here as you do with your Internet browser: **Command-[** (left bracket) to go back and **Command-]** (right bracket) to go forward.

TIP

Open in Timeline

Any clip in the Event Browser or in a project can be opened in the Timeline using the shortcut menu. Here, you have access to the video and audio of the clip in separate tracks. You can change the sync by sliding the audio independently in subframe amounts, or you can switch tracks on and off as needed.

RECORDING AUDIO

Final Cut Pro has a very simplified audio recording feature that allows you to record directly into your project. It allows you to record narration or other audio tracks directly to your hard drive while playing back your project. This recording is most useful for making *scratch tracks*, test narrations used to try out pacing and content with pictures. It could be used for final recording or emergency dialogue replacement, although you'd probably want to isolate the computer and other extraneous sounds from the recording artist. Many people prefer to record narrations before beginning final editing so the picture and sound can be controlled more tightly. Others feel that recording to the picture allows for a more spontaneous delivery from the narrator.

You can access the tool under the **Window** menu by selecting **Record Audio**, which brings up the panel in Figure 7.44.

The first pop-up allows you to select the event your recording will get saved into, and the second lets you set the Input Device. Any connected USB microphone should work well. The classic standalone FireWire iSight camera mic and built-in computer audio mics on MacBook Pros and iMacs are also recognized. The last pop-up sets monitoring. Make sure you have headphones plugged in if you check on monitoring, or you'll create feedback. There is a slider to adjust the monitoring level.

The controls couldn't be simpler: There is a Gain slider under the Input Device to adjust the recording level, horizontal meters monitor your levels, and one red button starts recording and stops recording. You can also stop recording by pressing the spacebar.

When you start recording, the project starts to play back, and the recording begins immediately—no countdown, no pre-roll, no post-roll, as in previous versions of Final Cut. The recording is stored in the designated event and is added to the project, connected to the primary storyline beginning at the point the recording started.

FIGURE 7.44

Record Audio Panel.

TIP

Switch Applications While Recording

A nice feature of FCP's Record Audio function is that you can switch applications while you're recording, and the record will not stop. So you can start the recording in FCP, switch to Pages with **Command-Tab** to read your script, and then switch back to FCP to stop the recording.

TIP

Playback Levels

Don't be fooled by FCP's vertical audio meters. These meters display the playback levels; they do not show the recording level.

After recording, the clip appears in the Timeline as *Voiceover 1*. Subsequent recordings are numbered sequentially. The clip is also added to the event. If you want to rerecord, simply delete the clip in the project and try again. When you've finished your recording session, make sure you also go through your event and delete the recording you're not using. Deleting them from the project does not delete them from the event.

SUMMARY

In this lesson we looked at working with sound in Final Cut Pro. We covered working with split edits, cutting with sound, using meters, overlapping and cross-fading tracks, setting Loudness and Audio Enhancements, performing audio matching and surround panning, editing music, and using markers. We saw the power of roles and the use of the Timeline Index; plus, we introduced FCP's audio recording tool. Sound is often overlooked because it doesn't seem to be that important, but it is crucial to making a sequence appear professionally edited. In the next lesson, we'll look at working with transitions in Final Cut Pro.

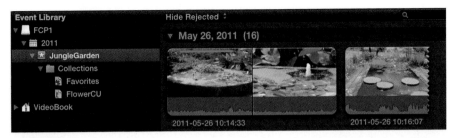

FIGURE 2.1

Event Library and Browser.

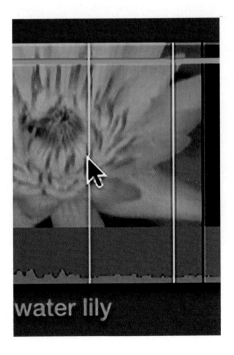

FIGURE 2.10

The Skimmer and Playhead.

FIGURE 3.13

Camera Import Window.

FIGURE 6.12

Media Limit Indicator.

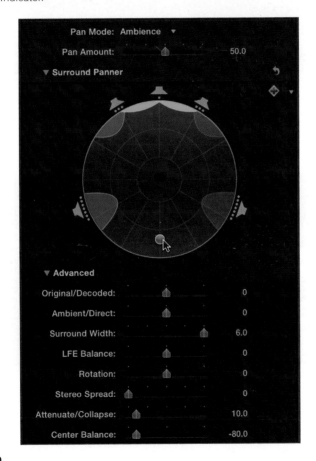

FIGURE 7.30

5.1 Surround Panner and Controls.

FIGURE 7.38

Roles Index.

FIGURE 8.18

Multiflip Transition.

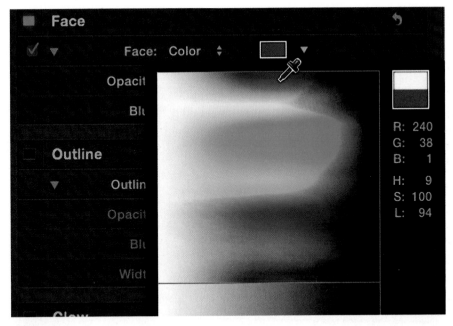

FIGURE 9.5

Face Controls.

FIGURE 10.17

Spot Image.

FIGURE 11.2

Histogram.

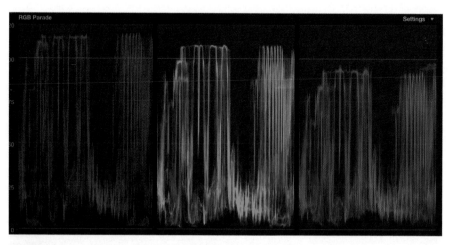

FIGURE 11.5

RGB Parade.

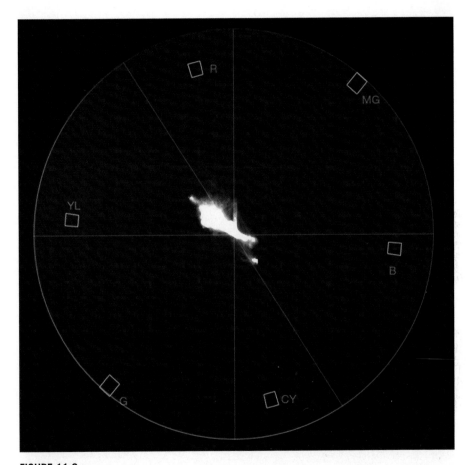

FIGURE 11.6

Vectorscope.

FIGURE 12.16

Default Ken Burns Effect.

FIGURE 13.2

Text with Color Dodge.

Adding Transitions

LESSON OUTLINE

Transitions can add life to a sequence, transform a difficult edit into something smoother, or give you a way to mark a change of time or place. The traditional grammar of film that audiences still accept is that "dissolves" denote small changes, and a fade to black followed by a fade from black marks a longer passage of time. With the introduction of digital effects, every imaginable movement or contortion of the image to replace one with another quickly became possible and was just as quickly applied everywhere, seemingly randomly, to every possible edit. They can be hideously inappropriate, garish, and ugly, but to each his own. Transitions can be used effectively, or they can look terribly hackneyed. Final Cut Pro gives you the option to do either or anything in between.

Let's look at the transitions FCP has to offer. There are quite a few of them, 88 to be exact. Some people seem to think that just because Apple put all those transitions in there, they have to use them all. Remember that most movies use only cuts and the occasional dissolve. Most film and television programs are cuts only, with a fade in at the beginning and a fade out at the commercial breaks. If there aren't enough transitions in FCP, or you need something different, more third-party transitions and effects are becoming available online from companies such as GenArts (www.genarts.com), who make the gorgeous Sapphire Edge effects.

In this lesson we'll look at adding transitions, controlling them, and adjusting the look of transitions using the available parameters.

> **NOTE**
>
> Motion 5
>
> One feature that's very different about these transitions and all the effects in FCP is that they are all created using Motion 5. Most of the transitions, titles, and effects in FCP can be altered by **Control**-clicking on the item in the media browser and selecting **Open copy in Motion**. It makes Motion 5 an extraordinarily powerful add-on to FCP, but one that requires a substantial learning curve.

LOADING THE LESSON

Let's begin by downloading the material you need on the hard drive of your computer, if you have not done so already. For this lesson we'll use the *FCP5.sparseimage*. You can download the material from the www.fcpxbook.com website.

1. Once the file is downloaded, copy or move it to the top level of your dedicated media drive.
2. Double-click the *FCP5.sparseimage* to mount the disk.
3. Launch the application.

> **NOTE**
>
> H.264
>
> The media files that accompany this book are heavily compressed H.264 files. Playing back this media requires a fast computer. If you have difficulty playing back the media, you should transcode it to proxy media as in previous lessons.

On the *FCP5* disk, you'll find an event called *Snow* that contains two Smart Collections: one of video clips and one with graphics. The Project Library also will show two projects that are on the image. The *Video* Smart Collection contains eight shots that we'll work with here.

APPLYING TRANSITIONS

In the Media Browser on the right side of the Toolbar is an hourglass button that will be familiar to iMovie users. It allows you to access the transitions available in the FCP Media Browser. You can also open the Transitions panel using **Window>Media Browser>Transitions**. In this panel you can see the available transitions with a preview that can be skimmed or played (see Figure 8.1). Note that skimming is enabled here, regardless of whether or not you have skimming turned on or off.

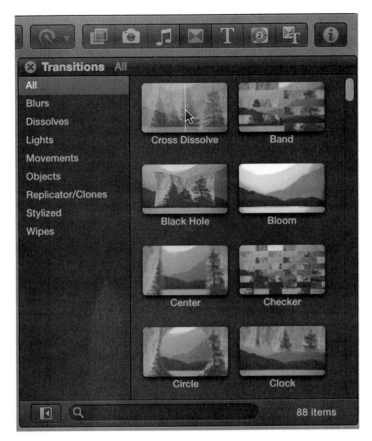

FIGURE 8.1

Transition Browser.

TIP

Keyboard Shortcut

While there is no keyboard shortcut to open the Transition Browser, you could easily map **Media Browser>Transitions** to a custom keyboard shortcut if you wish.

In the Transitions panel, you can see icons for each of the transitions. You can either see all the transitions by selecting **All** at the top of the list on the left or groups of them based on the transition type like **Dissolve** or **Wipe**. If you skim your pointer over the icons, you'll get a brief preview of the effect, which is displayed not only in the Transitions panel but also full size in the Viewer. This is really handy, and a new feature for Final Cut users, though not to iMovie users. I'd be very surprised if anyone has ever used all of these tools seriously on real projects and not just played around with them to try them out. We're going to try out some of them later in this lesson.

The default transition in FCP is the cross dissolve with a default duration of one second. But this is no simple cross dissolve: It has 11 different types of cross dissolve built into it.

You can apply a transition in Final Cut Pro in several different ways:

- Drag the transition you want from the Transitions panel and drop it on an edit point.
- Move the playhead to the edit point and select it with either of the bracket keys [or], or click on it with the Select tool. Then double-click the transition you want in the Transition Browser.
- **Control-**click on the edit point and select **Add Cross Dissolve** from the shortcut menu. This will apply transitions at both ends of the selected clip.
- Select the edit point and apply the default transition with the keyboard shortcut **Command-T**.

If you want to replace a transition in the Timeline with another one from the Transitions panel, select the one that's in the Timeline and double-click the transition you want to replace it with. Any duration change you made to the transition will be maintained.

If you wish, you can close the sidebar with the group list by clicking the word **All** to the right of Transitions, or by clicking the small triangle button in the lower right of the panel. Notice there is also a search box at the bottom of the Transition Browser; it's at the bottom of each panel in the Media Browser.

> **NOTE**
>
> Transitions on Both Ends
>
> If you select both sides of an edit point with the backslash key (\), the transition may get applied not only to the edit point you've selected, but also to the edit points on the ends of the two adjacent clips. This behavior is inconsistent at this time.

Available Media

Before we apply any transitions, we need to make sure our Preferences are set correctly. In **Preferences (Command-,)** under the **Final Cut Pro** menu, you should make sure that the **Apply transitions using** pop-up is set to **Available Media** in the Editing panel. You should always ensure that the clips you edit into the project have extra media called *handles* that allow the clips to overlap and create the transition.

Just to see what happens, let's set the transition preference on **Full Overlap**. Before we begin, though, I suggest duplicating the *Snow* project:

1. Select the *Snow* project in the Project Library on the *FCP5* disk and use **Command-D** to duplicate it.
2. Leave **Duplicate Project Only** checked and uncheck **Include Render Files**.

3. Duplicate the project to your system drive or your media drive outside the *FCP5* disk. This will preserve the original for you to return to if necessary.
4. You can rename it to something like *Snow copy*.

Now that we've got the housekeeping out of the way, we're ready to work with our transitions:

1. Start by opening the project *Snow copy* from the Project Library.
2. Select the five clips in the project with **Command-A**. The project is around 13 seconds long; it says so at the bottom of the FCP window.
3. Press **Command-T** to apply the default transition. The project now isn't even nine and half seconds long.
4. Undo that with **Command-Z** and set the transition preference to **Available Media**.
5. Again, select all the clips and apply the default transition. Transitions are applied, but the project remains the same duration.

Media has been taken off both sides of every shot to create the transition when the preference is set to Full Overlap. If you leave sufficient handles for the transition in the clips you edit into the project, you can use Available Media. These clips were edited into the Timeline with handles.

NOTE

Transition from Black

If you've applied cross dissolves to all the clips, you've also added a fade-in at the beginning and a fade-out at the end. These fade to black; they do not fade to transparency. If, for some reason, you need to fade to transparency, **Control**-click on the transition and uncheck **Transition to/from Black**.

Let's see what happens if clips that don't have handles are edited into the project:

1. Select the last two clips in the *Video* Smart Collection—*ryan mike n tyler* and *tracking*—and press the **E** key to append them to the end of project.
2. In the Timeline panel, set the project to Zoom to fit with **Shift-Z** so you can comfortably see the two new clips.
3. Click on the edit points between the last two clips. Notice the selection is red instead of the usual yellow, indicating you're on the limit of the available media. It doesn't matter which side of the edit you select; both will be red.
4. Double-click the edit point to open the Precision Editor and notice that the clips are simply butted against each other and have no overlap (see Figure 8.2).
5. Close the Precision Editor with the **Escape** key.
6. Apply the default transition anyway by pressing **Command-T**.

FIGURE 8.2

No Available Media in the Precision Editor.

A warning dialog (Figure 8.3) appears, asking if you want to use Full Overlap to do the transition. Because you have no extra media available on either side of the edit point, you can either shorten the clips manually or accept the dialog and allow it to use Full Overlap on the transition. If you do, the project will be shortened by one second, but if you really want the transition, you have no choice.

FIGURE 8.3

Insufficient Media Warning Dialog.

> **NOTE**
>
> Rendering
>
> As you start adding transitions, you'll notice an orange bar appearing in the Timeline above the transition. This bar indicates that the material needs to be rendered. If you have Background Rendering turned on in Preferences, after a brief pause, rendering will begin. If you do anything in the application, rendering will pause and then resume again seamlessly when it can. If you are working with proxy media and switch to optimized or original media, FCP will have to rerender. It keeps separate render files for each set of media. Once the media that needs to be rendered is processed for different formats, you can switch between them without needing to rerender.

ADJUSTING TRANSITIONS AND CLIPS

Before we look at the transitions themselves, let's look at how we can control the duration of the transitions and control the edit points that are now covered with the transitions.

To change the duration of the transition, to make it longer or shorter than the default, you can do the following:

- Drag one edge of the transition, pulling it in or out, as in Figure 8.4.
- Select the transition, press **Control-D** for the duration control in the Dashboard, and type in a new value.

To edit the shots underlying the transition, you have three grab points on the bar at the top of the transition. The ones on either end allow you to ripple the underlying shots, as in Figure 8.5. The one on the right grabs the end of the outgoing shot, the A side, and lets you ripple that shot, shortening and lengthening it as you like. The one on the left selects the beginning of the incoming shot, the B side, and lets you ripple that shot, again shortening and lengthening it as you like. The grab point in the middle of the transition (see Figure 8.6) lets you roll the edit up and down the Timeline. No need to select the Trim tool to roll the edit point here.

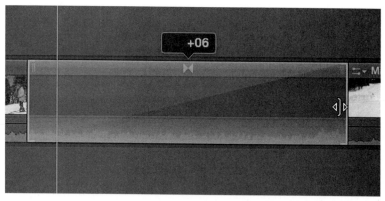

FIGURE 8.4

Dragging a Transition Longer.

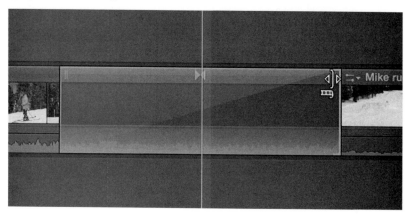

FIGURE 8.5

Rippling the Underlying Shot.

You can also trim the edits under the transitions numerically and by nudging them. If you use the Up and Down arrow keys to move between the edits, you'll notice that on most of them, the playhead makes three stops at each transition. It will stop at the beginning of the transition, at the middle of the transition, and at the end of the transition. This allows you to select those points on the transition handle just as if you were grabbing them with the mouse. Let's see how to do this:

1. Use the **Up** and **Down** arrows (or **semicolon** and **apostrophe** keys, whichever you prefer) to move the playhead to the left edge of the transition between *hand held trash can* and *Kevin 3 tables*.
2. Press either the left or the right bracket keys ([or]) to select the edge of the transition and ripple the incoming shot.

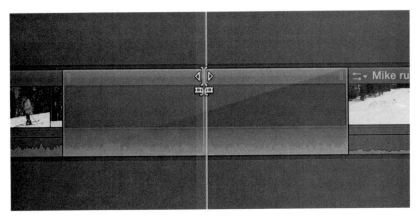

FIGURE 8.6

Rolling the Edit Under the Transition.

FIGURE 8.7

Rippling Edit Numerically.

3. To shorten the shot, use the **period** key to nudge it; to lengthen the shot, use the **comma** key. If you had the right edge selected, you would use the opposite keys: **period** to make it longer and **comma** to make it shorter. You can also use **Shift-comma** and **Shift-period** to ripple trim the shots in 10-frame increments.
4. Use the keyboard **semicolon** or **apostrophe** to move the playhead to the middle of the transition and press the **backslash** key to select the edit point under the transition in Roll mode.
5. Use the **period** (or **Shift-period**) to nudge the edit later in time and **comma** (or **Shift-comma**) to nudge the edit, rolling it earlier in time.
6. To do the edits numerically, simply select whichever part of the transition you want—beginning, middle, or end—and with the yellow selection indicator showing, simply type either + or – whatever value you want (see Figure 8.7). Plus will move the end farther along the Timeline, moving it to the right and later in time, and minus will move it farther back along the Timeline, moving it to the left and earlier in time.

TRANSITIONS AND CONNECTED CLIPS

Transitions that are applied to Connected Clips behave a little differently. Transitions can be applied only to clips that are in a storyline. For the primary storyline, as in this project, it's not a problem, but what happens if you've added Connected Clips? It's not a problem. How do you do a transition into the Connected Clips, and how do you do a transition between Connected Clips? Applying a transition to Connected Clips will automatically create an additional storyline:

1. Duplicate the project called *SnowConnected* onto your system drive—just the project, nothing else.
2. Open the duplicate project. It has one clip on the primary and three clips connected to it.
3. Drag-select or **Shift**-select the three Connected Clips and press **Command-T**.

    All of the edits have cross dissolves applied and are converted to a storyline, pinned at the first frame (see Figure 8.8). You need to be careful, however, if you start to apply transitions to individual clips.
4. Undo the applied cross dissolves with **Command-Z.**
5. Select the edit point at the start of the first Connected Clip and press **Command-T**. A storyline is created that extends over the single clip (see Figure 8.9).

FIGURE 8.8

Additional Storyline with Transitions.

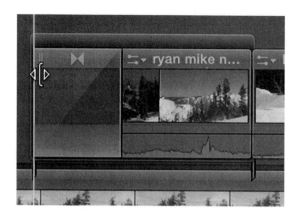

FIGURE 8.9

Single Clip Storyline.

FIGURE 8.10

Additional Storyline.

FIGURE 8.11

Break Apart Warning.

6. Select the edit on the last of the Connected Clips and add a cross dissolve to that. A separate storyline is created for that one clip.
7. In this instance it might be better to undo the transitions, select all the clips, and convert them to a single, additional storyline with **Command-G** (see Figure 8.10).

The continuous storyline is treated as a single unit, pinned on the first frame, and can be moved as a unit. It can also be broken apart at any time. If you do break apart the additional storyline, the transitions will be lost, but not until after you get the warning shown in Figure 8.11.

NOTE

Transition from Transparency

I noted earlier that the fade-out and fade-in at the beginning and end of the primary storyline defaults to fading from black; in the case of additional storylines, the fades at the beginning and end default to fade from transparency, not from black, which is probably what you want.

TRANSITIONS AND AUDIO

Whenever you have video and audio together, as we do here in the Timeline, and you apply a transition, FCP always applies an audio cross fade as well. For a great deal of material, this is fine and works well, and that's generally the way this application is designed—to do what's most common as the default behavior. Unfortunately, when audio and video are concerned, you seldom cut them together, so you often have different priorities for where cross fades are placed.

If you want to apply a video transition and not an audio cross fade, you have to expand the audio from the video. Let's start by looking at the transitions the way they are at the moment:

1. Open the *Snow copy* project again and zoom into the area around one of the transitions.
2. Select the two adjacent clips around the transition and press **Control-S** to expand them. What you'll see is shown in Figure 8.12: two tracks of audio overlapping, each with a fade applied.
3. If you don't want the audio cross fade, select the transition and delete it. Now there is no transition at all.
4. Click on the video edit to select it or press a **bracket** key and then apply the transition using the keyboard shortcut or double-clicking the one you want. Now only a video transition is applied with no audio cross fade, as in Figure 8.13.

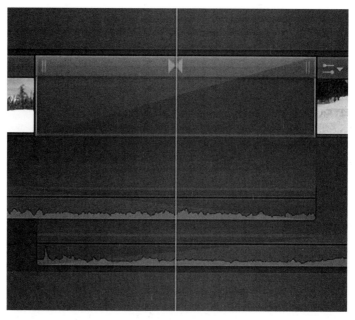

FIGURE 8.12

Video and Audio Transition.

FIGURE 8.13

Video Transition Without Audio Cross Fade.

You get this effect because the clip is expanded. While it's expanded, the cross dissolve can be applied separately to the video. If you want to apply an audio fade only, which you very often do while keeping the edit a cut, you have to do that manually, although there is an alternative in the note titled "Cross Fade Transitions."

1. With the clips collapsed, start by deleting the transition that's in the Timeline.
2. Expand the clips with **Control-S** so the audio moves down and can be adjusted separately.
3. Drag the two audio tracks so they overlap, as in Figure 8.14. The tracks of course slide out of the way of each other.
4. Grab the fade handle at the beginning of each clip to pull out the audio cross fade.
5. Select the clips and press **Control-S** to collapse the tracks again.

Another way to make the audio cross fade is quicker perhaps but more destructive, in that you have to separate the audio from the video into connected audio clips and convert them to an audio storyline. To see how to do this, undo what you did previously with the fade handles. You can either collapse the audio again or not; it doesn't matter. Then do this:

1. To make the cross fades, select the clips in the Timeline and use **Clip>Detach Audio** or **Shift-Control-S**. Do NOT use **Break Apart Clip Items**. This will appear to do the same thing but will have a very different outcome.
2. With the audio detached, select the audio portions of the clip and press **Command-T**. A cross fade will be applied between the audio clips, and they will be converted to an audio storyline, with one connection point at the head.

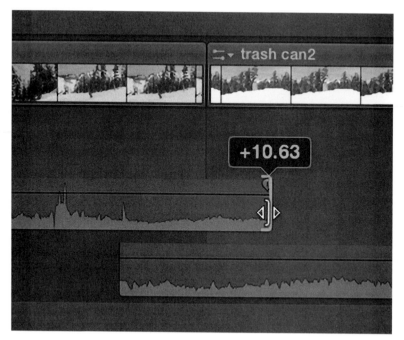

FIGURE 8.14

Overlapping Audio for Cross Fades.

3. There is no reattach function in FCP, but you could simplify your project time-line by selecting the video and audio and converting the selection into a single Compound Clip with **Option-G**.

If you broke apart the audio from the video rather than detached it, the audio would fade out and fade up at the beginning and end of every clip. The result would not be a smooth cross fade between the sounds, but a constant rising and lowering of the audio levels. That might be what you want in some circumstances, but probably not in most.

NOTE

Cross Fade Transitions

There aren't any cross fade transitions in FCP the way there were in FCP7 and earlier versions and in FCE. The cross fades are applied either with the video transition or by detaching the audio (NOT by breaking apart the audio) and using the same cross dissolve without the video selected. Alex Gollner has created many useful tools for FCP users, one of which is an audio cross fade transition that can be found at http://alex4d.wordpress .com/2011/07/11/fcpx-transition-sound-only/.

Copying Transitions

Sometimes you've applied a customized transition and you want to apply it again. There are basically two ways to do this; these steps show how:

1. In the *Snow copy* project remove a few transitions but leave one in place.
2. Select the transition in the Timeline and copy it.
3. Move to another edit point and use the **bracket** key to select it, or click on it to select it.
4. Use **Command-V** to paste the transition onto the edit point. You'll have to select and repeat this for multiple edit points.
5. Undo the paste of the transition, and we'll try the second method.
6. Hold the **Option** key; then grab the transition and drag it to another edit point. I think this is the simplest method if you want to apply the transition to multiple edit points.

TRANSITION BROWSER

With the basic cross dissolve applied, we'll start looking at our transition controls for the default transition:

1. With the transition selected in the Timeline, open the Inspector by clicking the **i** button or pressing **Command-4** (see Figure 8.15).
2. Change the Look pop-up to try different types of cross dissolve transitions.
3. You can also try different types of audio cross fades although the default +3dB probably works best with overlapping audio. Every transition has these audio controls.

FIGURE 8.15

Transition Inspector.

Many of the transitions have multiple controls. The **Directional** blur transition can be controlled by a dial in the Inspector, but it also has an on-screen HUD in the Viewer to control the direction and intensity of blur (see Figure 8.16). Turning the control vane will change the direction; pulling the vane out or pushing it in will increase or decrease the amount of blur. The blur amount can only be controlled on the screen, not in the Inspector. The **Radial** blur has an on-screen target that you can drag wherever you want to position the center of the blur effect. The **Zoom** blur transition has a similar control. Try moving the zoom point to one of the corners of the screen to create an interesting effect. The **Zoom & Pan** has two controls, the green one for the start and the red one for the end, allowing the zoom effect to move across the screen in whatever direction you want.

The **Fade to Color** dissolve fades to black, but you get to pick whatever color you want. If you want the ever-popular white flash transition, the **Flash** transition in the Lights group might be better for you. The Fade to Color transition has some interesting controls in the Inspector (see Figure 8.17). You can change the Midpoint,

FIGURE 8.16

Directional Blur Controls.

FIGURE 8.17

Fade to Color Controls.

allowing one side to fade in or out more quickly than the other. You also can control how long the color holds in the middle, so you could have a quick fade to black, a long hold, and then a quick fade up from black, or any number of different combinations created in conjunction with the Midpoint slider.

The Lights group of transitions have no video controls at all, though some, like **Bloom** and **Light Noise**, produce interesting effects, as does **Drop** in the Movements group. In the same group, **Multiflip** allows you to put a colored background or even a third image or video file underneath the two transitioning clips by dropping the item in the Image Well in the Inspector (see Figure 8.18, also in the color section). **Page Curl** has a great many controls, including an on-screen HUD that will allow the image to curl in one direction and peel back in the opposite direction, as in Figure 8.19. It really does give a great variety of control to this very traditional transition. You can also add a color to the backing of the image and change the Direction to either open (peeling

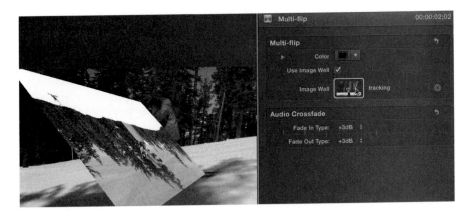

FIGURE 8.18

Multiflip Transition.

FIGURE 8.19

Page Curl Transition.

off the screen) or close (peeling onto the screen). Also in the Movements group is a useful transition that's easily missed: the classic Push Slide. It's hidden in the **Slide** transition. In the transition's controls, a pop-up lets you select between **Slide In**, **Slide Out**, **Slide Push**, and **Slide Swap**. You can also control the direction of the slide with a pop-up or even create a custom direction with on-screen controls.

The Objects group is a pretty cheesy collection, including theatrical red **Curtains**; tumbling **Leaves**, whose colors can be seasonally changed and customized; and the falling **Veil**, which can be colorized to suit your wedding theme.

The two transitions in Replicator/Clones have interesting Timeline controls that allow you to put multiple shot onto the screen. We'll look at **Clone Spin**, which has nine separate images that sweep in front of the viewer and go from the original shot to the new shot. With the transition selected, nine marker flags appear on the storyline (see Figure 8.20) that can be dragged anywhere and rearranged in any order that you want in the project, allowing you to select the image that appears in the moving video wall.

Some of the other transitions in the Stylized group, such as **Pan Far Right**, have these Timeline frame selection controls as well. This group is further subdivided into clusters of transitions that are parts of Themes, which are groups of graphical content including transitions and titles based on types of content: News, Sports, Bulletin Board, Cinema, Comic Book, Event, Nature, and others. Here, you have access to the appropriate transitions for the themes you're working in. These can also be reached in the Themes Browser, where you can open up both the transitions and their paired titles (see Figure 8.21).

Finally, the Wipes group has a number of standard wipes: **Circle**, **Clock**, **Letter X**, and others. The **Gradient Image** is a powerful transition, as it is fully customizable by the use of other images. Here's how to use it:

1. Select one of the edits in the *Snow copy* project and double-click the **Gradient Image** transition to apply it.
2. Select the transition in the Timeline and make sure the Inspector is open so you can see the Image Well in the controls.
3. From the *Graphic* Smart Collection, drag the *Gradient* image into the Image Well. This will produce the effect in Figure 8.22.

FIGURE 8.20

Clone Spin Timeline Frame Controls.

FIGURE 8.21

Themes Browser.

FIGURE 8.22

Gradient Image Transition.

Any grayscale image can be used to create the effect. You could even put video in the well, but only its luminance values are used. It has the potential for creating innovative and unique effects.

SUMMARY

That's it for transitions! We looked at how to apply them, the importance of handles for an Available Media transition, and the difference from Full Overlap. After applying the transition, we adjusted them in the Timeline. We also looked at applied transitions to Connected Clips and the way transitions work with audio, automatically creating cross fades, and how to avoid this effect when necessary. We looked at the Transition Browser and looked at some key transitions. We also saw how to customize them, copy them, and replace them. Everybody has his favorite transitions. Mine are fairly simple—mostly cross dissolves, occasionally a slide or wipe. Most of them I have never used and never will. Many probably should never be used, and you'll probably never see most of them. Next, we'll look at some of the huge variety of text and titling options that are available in FCP.

Adding Titles and Still Images

9

LESSON OUTLINE

Every program is enhanced with graphics, whether they are a simple opening title and closing credits or elaborate motion-graphics sequences illuminating some obscure point that can best be expressed in animation. This could be simply a map with a path snaking across it or a full-scale 3D animation explaining the details of how an airplane is built. Obviously, the latter is beyond the scope of both this book and of Final Cut Pro alone, but many simpler graphics can be created easily within FCP. More advanced motion graphics can be done in FCP's companion application, Motion, but that would be the subject for another book. In this lesson, we look at typical titling problems and how to deal with them. As always, we begin by loading the project.

SETTING UP THE PROJECT

For this lesson we'll use the *FCP5.sparseimage* as in the preceding lesson. If you haven't downloaded the material, or the project or event has become messed up, download it again from the www.fcpxbook.com website.

1. Once the file is downloaded, copy or move it to the top level of your dedicated media drive.
2. Double-click the *FCP5.sparseimage* to mount the disk.
3. Launch the application.

> **NOTE**
>
> H.264
>
> The media files that accompany this book are heavily compressed H.264 files. Playing back this media requires a fast computer. If you have difficulty playing back the media, you should transcode it to proxy media as in previous lessons.

The sparse disk has a couple of projects in its Project Library that we've already used.

To begin, we'll start by making a new project:

1. In the Project Library, select your system drive or, better yet, your media drive, and press **Command-N**.
2. Name the new project *Titles* and set the *Snow* event as the default event. The empty project will open in the Timeline.

TITLE BROWSER

Let's look at FCP's titling options. To access the Title Browser, click the **T** button in the Toolbar portion of the media browser (see Figure 9.1). You can open it from **Window>Media Browser>Titles**. There are no fewer than a staggering 159 text tool options. If you know the name of the text option you want, there is a search box at the bottom. On the left, the titles are broken in groups, the two primary ones being Build In/Out and Lower Third.

Build In/Out

What is Build In/Out? you're probably asking. You need to understand that all the titles and effects and transitions in FCP are created using Motion 5. Almost all of them involve some sort of animation to bring the text onto the screen and to take the text off the screen. The Build In function moves or fades the text into place. The Build Out function takes it away.

To preview the animations, you can either skim over it with the pointer or, better yet, click on one like Assembler and press the spacebar. The preview will play in looped

FIGURE 9.1

Title Browser.

mode in the Viewer. There are 45 text animations. One doesn't have any animation. Some are fairly simple, such as Fade, which uses a blur wipe to affect the text; Fold, which has a drop zone and unwraps itself; and the ever popular Far, Far Away, which was taken from *Star Wars*. Some, such as Pointer List and Slide Reveal, have multiple lines of text, and some, such as Ornate, should probably never be used with video.

If you scroll down to the bottom of the group, you'll see some titles that are under headings, like Bulletin Board, Comic Book, and Sports. Though you can use these titles anywhere, they are specifically designed to be a part of a theme. We'll see themes later.

Before you go looking for a basic text tool, I'll just tell you: It's called Custom. It's the only text tool that is just on the screen, one line of text in the center of the image. It's called Custom because it has the most controls and can be used to create custom build-in and build-out animations. Despite its custom features, it can be used as a basic text tool. Before we apply the text, let's add some video to our project:

1. In the *Video* Keyword Collection in the *Snow* event, find the *tracking* shot and append it into the Time with the **E** key.

2. Switch to the Timeline with **Command-2** and set the video to fit the window by pressing **Shift-Z**.

3. Put the playhead at the two-second mark in the project and double-click the **Custom** title in the Titles Browser.

The title block appears in the project Timeline in purple connected to the video. This particular title is five seconds long. Others are different lengths, depending on their animations. You can do this to set the duration of the title:

1. Select the title in the project and press **Control-D** for duration or double-click the **Dashboard**.

2. Type in whatever time value you want—any value that's less than 24 hours anyway.

To change the text, double-click the word *Title* in the Viewer to select it, just as you would in a word processor, and type in your title. I typed in the word *SNOW*.

TIP

Background

If you place text over nothing in the Timeline, the blackness you see in the Viewer behind the clip is the emptiness of space. To make it a little easier to see some of the text types, go to user **Preferences,** and in the **Playback** tab, set **Player Background** to **Checkerboard**.

The title itself is seen in the Viewer on top of the picture (see Figure 9.2). The controls for the text are in the Inspector. It's important that the text be selected because this tool works very much like a word processor: You have to select something to change it. Because each line of text, each character, can be completely different—a different font, color, size—you have to make sure the text you want to change

FIGURE 9.2

Title in the Viewer.

is selected. You can select it by double-clicking it in the Viewer or drag-selecting individual glyphs. You can also select the items in Text tab.

TIP

Setting Title Position

If you set In and Out points to create a selection range in the Timeline and then double-click a title in the Title Browser, the new title will take its duration from the selected range and not use the default length.

You can also insert text into the Timeline by dragging to an insertion point between two clips, or by selecting the text in the Titles Browser and pressing **W**. You cannot use the Insert edit button as that always inserts from the Event Browser.

TEXT CONTROLS

There are two sets of controls for **Title** and for **Text**. Title holds the parameters published from Motion, whereas the Text tab gives you access to the common text-style functions you can use in FCP (see Figure 9.3). Let's start there:

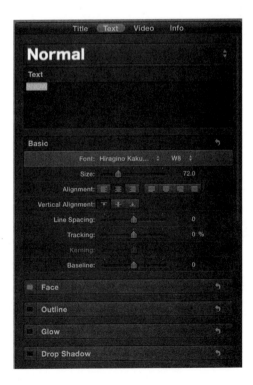

FIGURE 9.3

Text Controls.

1. You can change the text either by selecting it in the Inspector's Text tab or by double-clicking it right in the Viewer.
2. To change the **Font,** click on the Helvetica font and a pop-up appears with all your installed fonts. As you scroll through it, the text font will change in the Viewer. If you know the first letter of the desired font, type it to get to that letter group instantly.
3. The **Size** slider controls the point size of the text, and **Alignment** offers options for both alignment and text justification.
4. **Line Spacing** adjusts the gap between multiple lines of text.
5. **Tracking** controls the overall spacing between letters, while **Kerning** adjusts the spacing between individual letter pairs.
6. **Baseline** is useful if you have a single line of text, which normally sits on the center letter. Large font sizes appear high in the frame rather than centered. Adjusting the Baseline will let you shift the text so the words are centered on the screen, although usually having the text sit a little above center works well. Pull the Baseline slider to the left to lower the text.

Those are the Basic controls that can all be reset with the hooked arrow button opposite Basic.

TEXT STYLES

There are separate controls for Face, Outline, Glow, and Drop Shadow, which you have to double-click, or you can click on the hidden **Show** button to reveal them (see Figure 9.4). Aside from Face, each has to be activated with the blue LED checkbox on the left side of the control.

Face is the text color or gradient or texture. To the right of the color swatch is a disclosure triangle, which, brings up a color HUD, as in Figure 9.5 (also in the color section). Skim over the color HUD, and the color of the text will change in the Viewer. An Opacity slider and a Blur slider allow you to adjust the look and create interesting compositing effects.

Outline has similar color and Opacity and Blur controls as Face, but in addition it has the Width slider (see Figure 9.6). It's always a good idea to add either an outline or a drop shadow or both to text over video.

TIP

Adjusting Values

If you double-click in a value box such as for Opacity, you can type in a specific number. But even better, with the value selected, you can use the scroll wheel on your mouse to raise and lower the values and watch the effect in the Viewer over your video.

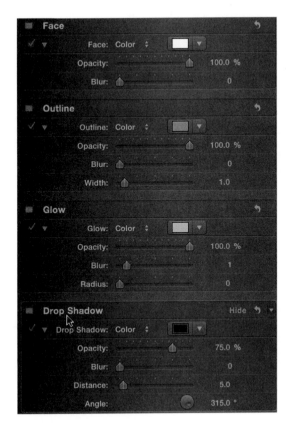

FIGURE 9.4

Style Controls.

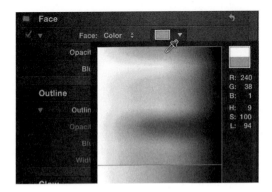

FIGURE 9.5

Face Controls.

FIGURE 9.6

Outline Controls.

For Glow to work effectively, you need to increase not only the Radius, but also the Blur value. Otherwise, the blur is hidden by the text, assuming the face is not transparent.

Even if you don't use an outline or glow, it's always a good idea to add a little **Drop Shadow**, especially with white text over snow. The default Drop Shadow is fairly weak and thin and needs to have its distance value pushed up a bit to somewhere between 15 and 20, depending on the point size of the text. Be sure to experiment with shadow color, using either the standard color wheel when you click the color swatch or the drop-down color palette and eyedropper, which appears when you click the adjacent arrow. You can watch the colors change on the text in the Viewer.

To position the text on the screen, make sure the text is selected. In the Viewer, you'll see a Position Target in the center of the baseline (if the text is center aligned, as in Figure 9.7). Simply grab the target and drag it anywhere on the screen you want.

> **NOTE**
>
> Fonts and Size
>
> Not all fonts are equally good for video. You can't just pick something you fancy and hope it will work for you. One of the main problems with video is its interlacing. Many video formats, both standard definition and HD 1080i, are still interlaced. *Interlacing* means the video is made up of thin lines of information. Each line switches on and off 60 times per second (50 times per second in Europe and other countries). If you happen to place a thin horizontal line on your video that falls on one of those lines but not the adjacent line, that thin horizontal line will switch on and off at a very rapid rate, appearing to flicker. The problem with text is that a lot of fonts have thin horizontal lines called serifs, the little footer that some letters sit on or have at the top, like the *N* and *W* in Figure 9.8.

FIGURE 9.7

Position Target.

FIGURE 9.8

Serif Font.

Unless you're going to make text fairly large, it's best to avoid serif fonts. You should probably avoid small fonts as well. Video resolution is not very high—the print equivalent of 72dpi. You could read this book in 10-point type comfortably, but a 10-point line of text on television would be an illegible smear. I generally don't use font sizes smaller than 24 point and prefer to use something larger if possible, depending upon the font structure and clarity.

Most of the titling tools in FCP do not have word wrapping. That means that as you type the text, it will type right off the screen. For your video to wrap the text onto another line, you need to manually enter returns in your text as you type across the screen when you get to the edge of the Safe Title Area. Televisions have a mask on the edge that cuts off some of the displayed picture area. What you see in the Viewer is not what you get and can vary substantially from television to television. That is why the Viewer can show a Safe Action Area (SAA) and a

FIGURE 9.9

Title/Action Safe.

smaller area that is defined as the Safe Title Area (STA)—the marked boxes seen in Figure 9.9. These are turned on with the switch menu in the upper right of the Viewer. At the bottom of the menu, select Show Title/Action Safe Zones. What's within the SAA will appear on every television set. Because television tubes used to be curved, and some older ones still are, a smaller area was defined as the STA in which text could appear without distortion if viewed at an angle. Titles should remain, if possible, within the STA. This is not important for graphics destined only for web or computer display, but for anything that might be shown on a television within the course of its life, it would be best to maintain them. That said, you will often see titles that are well outside the STA and lying partially outside even the SAA. Even with modern digital 16:9 flat screens, observing STA and SAA is best practice.

> **TIP**
>
> Safe Title and HD
>
> Though the STA for standard definition TV doesn't appear in FCP when working in HD projects, generally it's a good idea to keep titles in a safe area for standard definition as many viewers zoom into HD to avoid letterboxing on their television sets. If you leave your titles out at the STA for HD they would get cut off. Allow about three times more area on left and right for "center cut," the area seen on a standard definition TV without letterboxing.

At the top of Text controls is a big pop-up menu marked Normal. If you click it, you will see a whole array of wild and crazy text styles to choose from (see Figure 9.10). Just

FIGURE 9.10

Style Preset.

because they're available doesn't mean you have to use them. One of the most useful features is the ability to save both the basic attributes (font, size, alignment, tracking), as well as style attributes (face, outline, shadow), and to save both of them separately. You can select which combination you want to save at the top of the Style menu. No need to lose your masterpiece of graphic design; you can use it again and again.

Another simple way to save and reuse a title is just to copy it and paste it from one location to another in your project. You can also **Option-**drag a title from one location to another to duplicate it. Either way will preserve the formatting and style, and then you simply have to change the text.

TITLE FADES

Though most titles have some kind of Build In/Out effect, even if it's a simple fade, Custom does not. The simplest way to fade it in and out is to use the Opacity tools—not the slider in the Text controls, but the Opacity fade handles in the Timeline. Here's how:

1. With the *Snow* title selected in the Timeline, click the badge in the upper-left corner and select **Show Video Animations** or press **Control-V**.

FIGURE 9.11

Opacity Fades in Video Animation Control.

2. At the bottom of the list of video animation controls, double-click on **Compositing: Opacity**.

3. In the upper corners of the Opacity graph, just as for audio in the Timeline, are fade handles you can pull in to add and adjust the rate of fade-in and fade-out of the title (see Figure 9.11).

4. Click the close button (**X**) in the upper right or press **Control-V** again to close the video animation pop-up.

5. Play your video to see your title fade in and fade out at the end.

To delete a title, simply select it in the Timeline and press **Delete**. Another great feature of FCP allows you to simply replace one title with another rather than delete it. Here's how:

1. Select the *Snow* title in the Timeline.

2. In the **Build In/Out** group, find the Energetic title.

3. Double-click **Energetic** to replace the Custom title that was in the Timeline.

The title in the project is replaced, but the text remains the same. The basic attributes and style conform to the new title style, but the word *SNOW* is retained. This makes it easy to try lots of different title styles and animations.

BUMPERS, OPENS, CREDITS, AND ELEMENTS

The Bumper/Opener group contains a number of full-screen titles or full screens with video. You use them like this:

1. In the Event Browser, find the *Kevin 3 tables* shot and append it into the project with the **E** key.

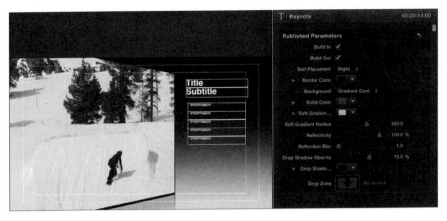

FIGURE 9.12

Keynote Controls.

2. In the Timeline, move the playhead back to the beginning of the *Kevin 3 tables* shot.
3. In **Bumper/Opener** group, find **Keynote** and double-click it.

The Keynote title appears with the video embedded in it, swinging open at the start to reveal title, subtitle, and bullet points. In the Title controls (see Figure 9.12), not to be confused with the Text controls, which adjust the text appearance, you can use controls that allow you to flip the text from the right to the left, controls to adjust the shading of the background gradient, and various other controls for the look of the title. Every title will have different controls depending on the look. Many of them are very well designed, and the adjustments make them very useful for a great deal of programming.

The Credits group includes a basic two-column scrolling title template as well as the basic movie trailer template that iMovie users will be familiar with from that application's movie trailers.

The Elements group has an instant replay tag in the upper left of the screen and that always-important speech or thought-bubble graphic.

LOWER THIRDS

The Lower Third group is an important group of title tools. A lower third is the graphic you often see near the bottom of the screen, such as those identifying a speaker or location that you always see in news broadcasts. There are 52 of them in FCP. From the silly Clouds to the popular Echo to the simple but effective Gradient – Edge, many of them are theme based. Let's replace the Energetic title that's over the *tracking* shot with the Gradient – Edge lower third:

1. Select the Energetic title in the Timeline and find **Gradient – Edge** in the Lower Third group.
2. Double-click **Gradient – Edge**.

FIGURE 9.13

Gradient – Edge Lower Third.

3. Select the title in the Timeline on top of the *tracking* shot and look at the Title Controls in the Inspector.

4. Double-click the first line of text in the Viewer and type in *TYLER SKIVINGTON* in caps.

5. Double-click the second line and type in *Bear Valley Park Crew.*

6. In the **Title** tab in the Inspector, push the Line 1 Size up to about 96 point and push the Line 2 Size to 70 point.

7. Increase the **Bar Width** to 150%. I also suggest lowering the **Bar Vertical Position** a little.

8. Click on the **Bar Color** swatch and choose a color from the wheel, or use its magnifying glass to pick a color, like the skier's jacket, from the frame.

9. Finally, you might want to go into the **Text** tab and add a little drop shadow to the text, making it fully opaque and moving it slightly farther from the text to help separate it from the background (see Figure 9.13).

Remember that the Apple Color Picker, unlike the drop-down palette color picker, can be used to save color swatches and is system wide, so it can be used in any application. Notice also that in the Text tab, you do not have text controls for multiple lines of text; the basic font and size controls for each line are in the Title tab.

All of the Lower Thirds will have slightly different controls.

FIND AND REPLACE

A great feature in FCP's titling tools is the ability to easily change text that perhaps was misspelled or had the wrong title. To do this:

1. Use **Find and Replace Title Text** from the bottom of the **Edit** menu.

2. In the dialog that appears, type *Tyler* in the Find box.

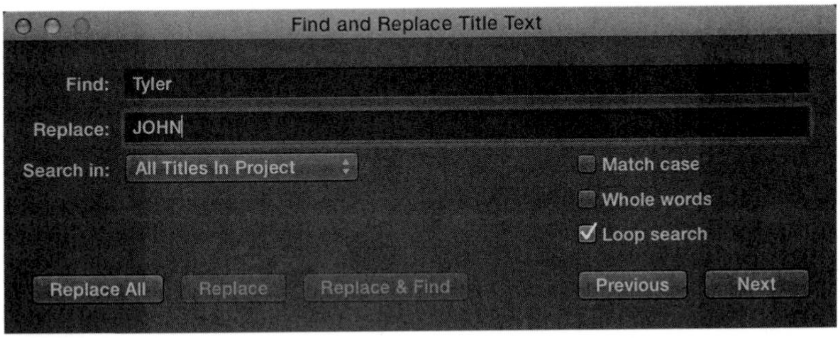

FIGURE 9.14

Find and Replace Title Text.

3. Type *JOHN* in the Replace box because we want the replacement text to be in all caps (see Figure 9.14).
4. Click the **Next** button to find the word and then click **Replace** to change it.
5. Close the box and Undo. (His name really is Tyler.)

If you have multiple instances, you could check each in turn or globally replace the word. By the way, in the Text tab of the Inspector, there is a built-in spell checker for the text that appears at the top of the screen. Incorrectly spelled words that aren't proper names will appear with squiggly red lines underneath them.

GENERATORS AND THEMES

Generators

To the right of the Title T button in the Toolbar is the button that accesses the Generators Browser. There are 26 generators, although third-party generators are available, many of them free—for example, the 12 Classic Generators from Ripple Training that include color bars, a countdown, and a grid.

The built-in generators include backgrounds, which have slight motion to them, like the rippling curtain and the shifting underwater light. Most have no published controls, except for curtain that allows you to change the color.

There are 12 textures, most of which have some kind of tint control. Stone lets you not only adjust the color, but also change the stone type into three kinds of concrete, or from Marble to Travertine or Slate or River Rocks.

Solids are solid colors. Custom allows you to pick any color you want, and Pastel lets you select from a narrow, pale palette. Vivid has primary colors, and Whites has different shades.

The Elements group has only four items, but they can be essential. Counting will create a counter for you in which you can set the start and end numbers so that you can count up or count down.

Timecode will generate timecode by reading the current time of the project and let you place it on top of the video.

The Shapes generator will put a shape over your video—from basic circles, squares, and rectangles to different types of stars, hearts, and arrows. You have full control over the appearance of the shape with fill color, outline, and drop-shadow controls. The Transform tools, which we'll see in Lesson 12, allow you to position and animate the shapes.

The final Generator is the Placeholder. This is a very powerful tool for not only putting a Placeholder in your project, but also actually building a storyboard of animatics if you wish. Let's see its controls:

1. In the Timeline, move the playhead to the end of the project and double-click the Placeholder in the Generators browser.
2. Select the Placeholder in the Timeline, and in the Inspector, click on the Generator tab at the top to see the controls.

You can set the size of the shot (long shot, medium-long shot, medium shot, close-up); the number of people from zero to five; the gender (male, female, or both); a string of different backgrounds (see Figure 9.15); weather type (cloudy, sunny, day, night); a checkbox to make it an interior; and a checkbox that lets you add text onto the screen (see Figure 9.16). You can add more shots simply by copying and pasting (or **Option-**dragging) the same Placeholder, changing the parameters for shot size and number of people, and changing the text. You can lay over the dialogue and

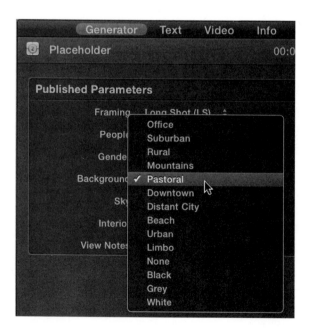

FIGURE 9.15

Placeholder Backgrounds.

Joan and her friends meet at the country house at night.

FIGURE 9.16

Placeholder Notes.

change the name of each Placeholder in the Info tab of the Inspector. When you're ready, and the movie is actually shot, you can use the Replace edit function to swap out the Placeholders with real video. This is great for all types of programs that have basic formulas, such as weddings and corporate videos and even commercials and narrative fiction.

Themes

Themes are a common concept to iMovie users. The idea is that there are collections of graphical tools, main titles, lower thirds, and transitions that work together as design elements, similar in color, shape, and animation. There are 15 groups of themes—from simple ones like Boxes, Tribute, and Cinema (see Figure 9.17) to complex ones like Bulletin Board, Comic Book, Scrapbook, and Photo Album that each include four transitions and eight title effects.

CUSTOM ANIMATION

One of the features of the Custom title that we applied earlier is its ability to be customized, hence its name. Let's apply the title and see how we can use the Build In/Out tools to bring the text onto the screen. Unlike keyframing, the build durations are set in Motion. Here, no keyframes are added because the animation goes to the title in its presentation setting, and from that, off the screen. First, let's set up the background and text:

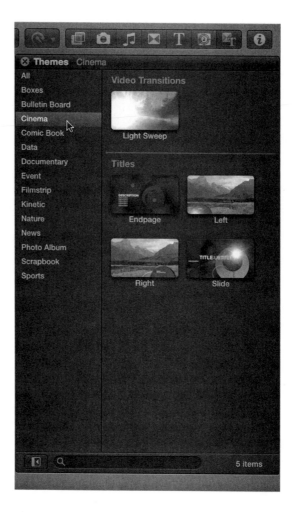

FIGURE 9.17

Cinema Theme.

1. Start by appending the *ryan mike n tyler* shot onto the end of the project.
2. Put the playhead in the Timeline at the start of the shot, find the Custom title in the Build In/Out group, and double-click it to put it on top of the video.
3. Select the text in the Viewer and change the words to read *Ryan Mike & Tyler*.
4. With the text selected in the **Text** tab of the Inspector, set the font to **Arial Black**, the **Size** to **100**, and turn on the **Drop Shadow**.
5. Push the **Drop Shadow Distance** up to **18** and set the **Opacity** to **100%**.

That's the text and video setup. You're now ready to create the custom Build In and Build Out in the Title tab of the Inspector. The controls look daunting (see Figure 9.18), but using them is easier than it appears. The controls are broken into

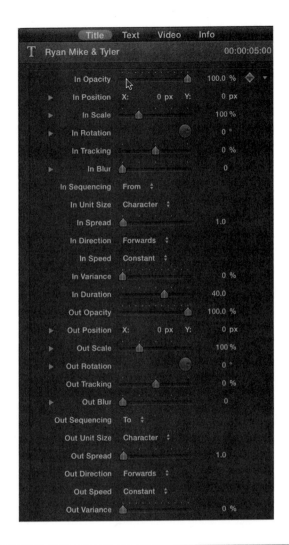

FIGURE 9.18

Custom Title Controls.

two equal halves—the first set from In Opacity to In Duration controls bringing the text onto the screen, and the same controls from Out Opacity to Out Duration controls taking the text off the screen. When you're setting up the In controls, you're setting up the start of the Build In; the end of the build is where the text is now—in the center of the screen. Similarly, when you're applying the Build Out, you're applying its end position; its start position for the Build Out is its current position in the center of the screen. The default durations for the animations are 40 frames; you can change that with the In Duration and Out Duration sliders. Let's build a little Custom animation:

1. In Unit, **Size** defaults to **Character**, which means each letter is handled separately. Set this to **Line**, which means the line of text will be handled as one unit.
2. Leave the **In Sequencing** set to **From**, so we'll be setting the position from which the animation will sequence.
3. Set **In Speed** to **Ease Out** so the animation will have some deceleration as it finishes.
4. Set the **In Blur** to **50**, which will start the image as very blurry.
5. Set the **In Tracking** to **100**, which will make it very wide.
6. Set the **In Scale** value to **215%**, which will make it very big.
7. Set the **In Position** Y value to **55**.
8. Play back your Build In animation.

Let's repeat the process, only in reverse. Let's do the Build Out:

1. Change the Out Unit **Size** to **Line**.
2. Change the **Out Speed** to **Ease In** so it accelerates.
3. Set the **Out Position** Y value to **–65** so the text goes off the bottom of the screen.
4. Set the **Out Blur** to **50** and the **Out Tracking** to **100**.

That's it! You've done your first Custom animation. We'll do more animations using keyframes in the lessons ahead.

COMPOUND TITLES

One of the strengths of Final Cut Pro has always been its ability to make complex layered titles. That is still the case in FCPX. Let's start with a new project, and we'll create a custom title composite that will give you some idea of what you can do with these title tools.

Text

1. Start by opening the Project Library (**Command-zero**) and selecting your system drive or media drive (not the disk image) to make a new project.
2. Name the new project *Compound Title*.
3. In the browser, find *Mike run 3 tables* and append it into the Timeline with **E**.
4. Play the clip till just after the skier lands his first jump. That's where we'll add the title.
5. In the Build In/Out group, find the Fade title and double-click it to add it to the Timeline on top of the picture.
6. In the **Title** tab of the Inspector, set the **In Size** to **Line** and the **Out Size** to **Line** as well.
7. Double-click the text in the Viewer to select it and type in *BEAR VALLEY* to replace *Title*.
8. You can use whatever font, color, or settings you want, but for this demonstration I'm going use the following settings:

Font	Arial Black
Point Size	120
Face	White
Outline	Black
Outline Width	4
Outline Blur	1.5
Drop Shadow Opacity	100
Drop Shadow Distance	16
Drop Shadow Blur	3

Texture

1. Click on the title in the Timeline to select it and then press the **X** key to make it a range selection.
2. From the **Generators** browser in the Textures group, find the Grunge texture and press the **Q** key to connect it to the primary storyline.
3. The texture is on top, so grab the title and drag it straight up so that it's on top of the texture. Hold the **Shift** key to constrain the vertical drag.
4. We'll look at cropping in more detail in Lesson 12, but for now select the texture in the Timeline and click the **Crop** button in the Viewer (see Figure 9.19).

FIGURE 9.19

Crop Button in the Viewer.

FIGURE 9.20

Crop Controls.

5. Grab the rectangular button in the middle at the top of the image and drag it down so it's just above the text.
6. Now grab the rectangular button in the middle at the bottom of the image and drag it up so it's just below the text (see Figure 9.20).
7. With the texture selected, go to the **Generator** tab of the Inspector and change the **Texture** to *texture 12*.
8. In the Video tab of the Inspector, set the **Opacity** to **60%**.

Because we have the text fading in, we want to fade in the background as well:

1. Click the Animation badge in the upper-left corner of the texture in the Timeline, and select **Show Video Animations** or use the keyboard shortcut **Control-V**.
2. Double-click **Opacity** to open up the graph area and drag the fade button to the right, just like the audio fade button we used previously in the Timeline.
3. Drag the fade button at the start to 1:10, which is 40 frames, the same as the title fade (see Figure 9.21).
4. Drag the fade button at the end to the same 1:10 to fade out the texture.
5. Close the Video Animations pop-up with **Control-V**.
6. To lengthen the two items, select them and press **Control-D** for the duration in the Dashboard.
7. Type in *+100* and press **Enter** to make the title elements one second longer.

Because the animations are not based on keyframes they are not linked to a specific time reference, the animations are still the same length.

Putting It All Together

1. Select both items in the Timeline by drag-selecting through them or **Command**-clicking them.
2. Use **File > New Compound Clip** or press the keyboard shortcut **Option-G** to convert the title elements into a single Compound Clip.
3. Select the Compound Clip, and in the Info tab of the Inspector, change the name to *Bear Valley*.
4. Append the *Kevin 3 tables* shot into the Timeline.
5. Select the *Bear Valley* Compound Clip in the Timeline, copy it, and paste it on top of the second shot.

FIGURE 9.21

Opacity Fade Buttons.

6. Rename the second Compound Clip *Jackson Hole*.
7. Double-click the second Compound Clip, which opens it into the Timeline, and change the text layer to read *JACKSON HOLE*.
8. Press the **Back** button or use the shortcut **Command-left bracket** to go back to the project and see the two title blocks in the Timeline, both with the same style and layout but with different text.

A Compound Clip is like a clip in a sequence. If you want to apply an effect to it or reposition the block—lower in the frame, for instance—you can do this without adjusting each layer individually. We'll look at applying effects in the next lesson, as well as animating images about the screen in Lesson 12.

TIP

Saving Compound Clips

If you want to keep a title for use in other projects, create a new Compound Clip in the Event Browser. Open the Compound Clip and edit your title into it, setting it to whatever style and text you want. The Compound Clip with the title will remain available for you to use in any project you have that has access to the event that holds it.

STILL IMAGES

Often you must work with still images, photographs, or graphics generated in a graphics application such as Photoshop or Pixelmator. What you should first know about working in these applications is that you should use only the RGB color space—no CMYK, no grayscale, and no indexed color. They don't translate to video.

The recent versions of Photoshop have presets for working in DV, NTSC and PAL, 4:3, as well as widescreen. You should use these presets whenever you're making a multilayer or transparent image for use with FCPX. For HDV material, use either the HDV 1440x1080 Anamorphic preset if you're working in 1080i or the HDV 720 preset if you're working in 720p.

You will not always work with images that conform to video formats. Sometimes your graphic will be much larger, one you might want to move around or to make it seem as if you're panning across the image or zooming in or out of the image. To do this, you need an image that's much greater in size than your video format. How much bigger depends on your creative needs, based on how much the image will need to be magnified. Magnification always requires more data, a larger frame size, to keep the image sharp.

When you import a graphic file into FCP and place it inside a project, FCP will scale the image to fit the dimensions of the format you're working in, scaling up if the image is smaller than the project and scaling down if it's larger. Here's how this works:

1. In the Smart Collection *Graphics,* select the image called *Sutton* and go to the **Info** tab of the Inspector. The image is 5191×1497 pixels at 60fps, which is the maximum frame rate FCP uses.
2. Click on the preview in the browser so that the user preference duration is selected; four seconds is the default. Then append the image into the project (see Figure 9.22). Because the aspect ratio is very different from the HD frame, the image is letterboxed, showing space at the top and bottom.
3. In the **Video** tab of the Inspector, you'll see that **Spatial Conform** is set to Fit. Change it to **Fill**. Then try **None**. With None set, the image is enlarged because the pixels in the image are spread out to their normal width.
4. If you have the Timeline image selected and press **Shift-T** for transform and zoom out a lot in the Viewer, you'll see the image's relation to the HD frame. It's much larger (see Figure 9.23).
5. Click in the preview to select the *Sutton2* image in the browser and append that into the project. Because the aspect ratio of the image is 4:3 rather than widescreen, pillarboxing will appear on the sides of the image.
6. Set the **Spatial Conform** to **None,** and you can see the image is quite a bit larger than the frame. Set it to **Fill** so the image fills the frame with no pillarboxing.
7. Finally, click in the preview to select the *Sutton3* image in the browser and append that into the project. This image will also be made to fit the frame.

FIGURE 9.22

Wide Image in the Viewer.

FIGURE 9.23

Wide Image in Relation to HD Frame.

8. If you set its **Spatial Conform** to **Fill**, the image will be scaled up further. Now set the **Spatial Conform** to **None**. The image is much smaller than the frame (see Figure 9.24). Once again, the system displays the image at the normal size, but there aren't enough pixels to fill the frame, much less provide additional space to support zooms.

FIGURE 9.24

Small Image in HD Frame.

TIP

Still Image Duration

If you drag a still image into a project from the Finder or the media browser, it comes in at the default four-second duration, or whatever the user preference is set to. If you select a still in the browser, the selection is four seconds, and that's what's edited into the Timeline. But if you drag a still from the browser to the Timeline without making a selection, or drag a group of still images from the browser into the project, they will all be 10 seconds in length. If you need to change the durations of the stills in the Timeline, simply select them, double-click the **Dashboard** for the duration dialog, type in a time value, and press the **Enter** key. All the selected stills will have their durations changed. You can also **Control**-click on the selected images and use **Change Duration**.

Because FCP is scaling the *Sutton3* image to fit inside the frame, it may not be very sharp. Unfortunately, there is no easy way to tell how much scaling is being applied to an image. Generally, it's not considered a good idea to scale up an image over about 110% to 115%. This image is clearly being scaled a lot further than that.

Resolution

For people who come from a print background, an important point to note is that video doesn't have a changeable resolution. It's not like print, where you can jam more and more pixels into an inch of space and make your print cleaner, clearer, and crisper. Pixels in video occupy a fixed space and have a fixed size, the equivalent of 72dpi in the print world, which happens to be the Macintosh screen resolution. Dots per inch are a printing concern. Forget about printing resolution. Think in terms of size: The more pixels, the bigger the picture, just the way digital still cameras work. Don't think you can make an image 720×480 at a high resolution such as 300dpi or 600dpi and be able to scale it up and move it around in FCPX. Certainly, you'll be able to scale it up, but it will look soft, and if you scale it far enough—to 300%, for instance—the image will start to show pixelization. FCP is good at hiding the defects by blurring and softening, but the results are not really as good as they should be and become painfully evident on large-screen projection. FCP is a video application and deals only with pixel numbers, not with dots per inch.

Scanners, on the other hand, generate lots of dots, and dots per inch can be considered roughly equivalent to pixels per inch. This is very handy for the person working in video. This means that you can scan a four-inch by three-inch image, which at 72dpi scan is a quite small image, but at, say, 300dpi or 600dpi, your scanner will produce thousands of pixels, which will translate into video as a very large image that remains sharp in the HD frame.

If your scanner can generate an image that's 3,840 pixels across, it's making an image twice as large as an HD video frame. The advantage of importing "rich scans" like this for video is that you can now move that very large image around on the screen and make it seem as if a camera is panning across the image. Or you can scale back the image, and it will look as if the camera is zooming back from a close point in the image. Or you can reverse the process and make it look as if the camera is zooming into the image, and the result will be as sharp as the original image allows.

When it comes to increasingly popular digital stills, the same rules apply, except scanners are irrelevant: You only need to see how large the original camera image measures in relation to your chosen video format. Today's multi-megapixel cameras can capture huge images, three or four times wider than the 1,920 pixels needed for HD 1080—and thus support zooming into the image three or four times, and often more.

NOTE

Layered Photoshop Files

Unlike in previous versions of Final Cut Pro and Express, in this version, multilayer Photoshop files are not treated as sequences or Compound Clips or projects. They simply appear as flattened images, although Photoshop and other transparencies are carried over to the application. If you want to work with layered Photoshop files for motion graphics purposes, you'll have to use Motion 5.

SUMMARY

In this lesson we looked at FCP's Title Browser, starting with its Build In/Out titles and the basic text controls and text styles. We learned how to do simple fades using the video animation controls. We worked with Lower Thirds and the Find Text and

Replace feature. We also saw Generators and Themes and created a simple Custom animation. We built a Compound Title Clip, adding text and textures, putting it together into a title composite. We also looked at how FCP handles still images and the issues for still image resolution. In the next lesson, we will look at the effects available in the application.

Adding Effects

In this lesson we look at and work with Final Cut's effects. FCP offers a great variety of excellent effects, including great color correction tools, which we'll look at in the next lesson. These effects rely heavily on the graphics card for processing. What this means is that on slower computers, you won't get very good performance with them, and on some marginal computers, they'll be practically impossible to work with.

Unlike transitions that go between clips, effects are applied to individual clips. In addition to the effects included with the application, other programmers are creating effects to add to Final Cut Pro. Be sure to check out companies such as GenArts and its amazing Sapphire Edge plug-ins.

While earlier lessons may have used concepts and tools that were familiar to iMovie users, as we go further into the book, we will use concepts and techniques that are more familiar to Final Cut users as well as users of other professional video applications.

SETTING UP THE PROJECT

For this lesson we'll use a couple of sparse images. We'll use *FCP5.sparseimage* from the preceding lesson as well as *FCP6.sparseimage*. If you haven't downloaded the material, or the project or event has become messed up, download it again from the www.fcpxbook.com website.

1. Once the file is downloaded, copy or move it to the top level of your dedicated media drive.
2. Double-click *FCP5.sparseimage* and *FCP6.sparseimage* to mount the disks.
3. Launch the application.

> **NOTE**
>
> H.264
>
> The media files that accompany this book are heavily compressed H.264 files. Playing back this media requires a fast computer. If you have difficulty playing back the media, you should transcode it to proxy media as in previous lessons.

You probably thought there were an awful lot of titles. Well, there are even more effects—207 of them in fact, more than half of them audio effects. Truth be told, some of these effects aren't very useful, but a lot of them let you do some pretty amazing things with video. In this lesson we'll look at some of the useful FCP effects.

There is a short project in the *FCP6* disk that we will use to apply some effects:

1. In the Project Library, select the *Effects* project on the *FCP6* disk and use **Command-D**.
2. Duplicate just the project to your system drive or your media drive, not to the *FCP6* disk.
3. Do not duplicate the associated events or render files.
4. It's a good idea to rename the duplicate something like *Effects copy*.

APPLYING AN EFFECT

Effects are accessed from the Effects Browser, which you can open by clicking the Effects button in the media browser in the Toolbar (see Figure 10.1) or by pressing **Command-5**. The effects are divided into two main groups: video and audio. Each has groups within it, like Basics and Blur in video, and Distortion and Echo in audio.

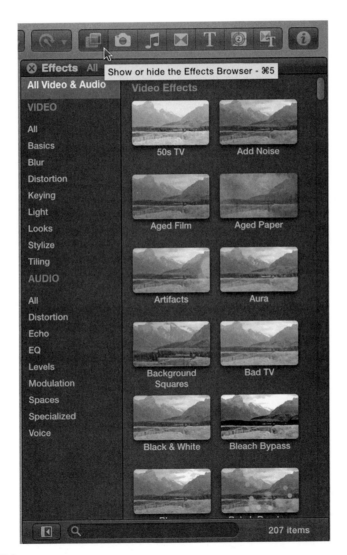

FIGURE 10.1

Effects Browser.

Effects are applied to clips in the Timeline; they cannot be applied to clips in the browser. Applying an effect in Final Cut Pro is really easy, but before you apply an effect, you can preview it with your video like this:

1. With the *Effects copy* project open, click on the first clip in the Timeline to select it.
2. In the Effects Browser, select the **Basics** group and skim over **Black & White**.
3. Try skimming over some of the others.
4. Click on the **Tint** effect and press the **spacebar**.

The effect not only will appear on the little icon in the Effects Browser, but also will appear full size in the Viewer. With the effect selected, the clip will play in Looped mode so you can see the effect.

TIP

Filmic Black & White

Here's a tip from Denver Riddle of Color Grading Central: In the Black & White effect, pull down the green slider under Color and push up the red slider to create a black-and-white image that more closely resembles black-and-white film response.

To apply an effect, you can either double-click on it or drag it onto a clip.

1. Go to the **Distortion** group of video effects and double-click **Background Squares** to apply it to the selected shot in the Timeline.
2. To access the effect controls, go to the Video tab of the Inspector and double-click the effect name (see Figure 10.2).

It's just as easy to remove an effect:

• Click on the effect name in the **Inspector** and press the **Delete** key.

You can also switch off the effect, while leaving it in place:

• Click the blue **LED** box next to the effect name to toggle the effect off and on.

This allows you to leave an effect in place while you toggle its effect on and off to see what it's doing to the picture.

FIGURE 10.2

Effects Controls in the Inspector.

Any number of effects can be added to a clip, but the order in which the effects are applied to the clip can be important. Try this exercise:

1. With the **Background Squares** effect applied to the first clip, make sure the clip is still selected in the Timeline.
2. In the **Blur** group of effects, find **Gaussian** and skim over it to see the result of the two effects applied together.
3. Double-click the **Gaussian** effect to apply it to the clip.
4. Push up the **Gaussian** amount to about **70** so it's very blurry. The Background Squares are hardly visible.
5. Drag the Gaussian effect in the Inspector so it's above the Background Squares and see the result in the Viewer, as in Figure 10.3.

Because the Gaussian blur is applied to the image before the Background Squares, the clip is blurred first, but the squares show sharp edges as they are applied after the blurring effect.

TIP

Adjusting Settings

In the Effects Browser if you **Option** skim a preview, you will see how the primary control can be adjusted and modified. As you skim with the **Option** key, adjusting the parameter, you will only see one frame, but when you release the Option key, you can skim the image to see the effect on the whole clip with the new settings. If you then apply the effect, it will be applied with the adjusted settings and not the default settings.

Copying and Pasting

Effects can also be copied and pasted from one clip to another:

1. Select the first clip, with the two effects applied to it, and copy it (**Command-C**).
2. Select the second and third clips in the Timeline and use **Edit>Paste Effects** or the keyboard shortcut (**Option-Command-V**) to paste the same effects with the same settings to those clips.

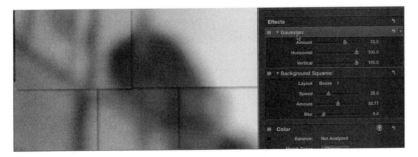

FIGURE 10.3

Viewer and Effect Ordering.

This is the easiest way to apply the same effect with the same settings to multiple clips. The ability to copy and paste effects applies not only to effects alone, but also to color correction and to transform properties that we will see in Lesson 12.

> **TIP**
>
> Removing Effects and Color Correction
>
> There is no remove effects function in FCP, but there is a really simple way to reset a clip or any number of clips, removing any effects and color corrections. Just select any unaffected clip (even in the Event Browser), copy it, select the clips you want to reset, and use **Option-Command-V** to paste "no-effects," which instantly removes effects and resets the clips.

Animating Effects

Many effect values can be animated just like audio parameters by adding keyframes so they can be altered over time. Let's do a simple animation of the blur value on the second clip in the Timeline. There are two ways to do this: either entirely in the Inspector or in the Timeline. We'll do both, starting with the Inspector:

1. Make sure the second clip is selected and that the playhead is at the beginning of the clip.
2. In the **Video** tab of the Inspector, move the **Gaussian Amount** down to zero.
3. To the right of the slider and the value box is a diamond button with a plus (see Figure 10.4). The button will go orange, indicating a keyframe has been applied.
4. Move the playhead forward one second in the Timeline and click the keyframe button again. You have to click the button to tell FCP you're changing the value.
5. Drag the **Amount** slider up to **70** or type that number in the value box. The image will get more blurry over one second.
6. Move forward one more second into the Timeline and click the keyframe button to add another keyframe. If you don't change the value, the effect will remain blurry for that one second.
7. Now go forward one more second and click the button again to add a fourth keyframe.
8. Drag the **Amount** slider down to **0** (zero). The image will now get blurry, hold blurry, and then go back to being clear.

FIGURE 10.4

Adding a Keyframe.

Let's do a similar animation to the first clip in the Timeline, but we'll do it entirely using the Timeline controls:

1. Select the first clip in the Timeline and apply the **Gaussian Blur** by double-clicking the effect in the Effect Browser.
2. Press **Control-V** to open the Video Animations control or click the badge in the upper-left corner of the clip.
3. With the controls open, double-click **Gaussian Amount**, or click the small arrow at the right, to reveal the keyframe graph (see Figure 10.5).
4. The **Amount** line should be 50% of the way up the graph. Drag the line all the way down to **0** (zero).
5. Hold the **Option** key and click on the line right at the start to add a keyframe. Zooming into the Timeline will make it easier.
6. Move the playhead forward about one second so you have a snapping guide for the pointer. Make sure Snapping is turned on with the **N** key.
7. Put the pointer on the line, and holding the **Option** key, click on the line to add a keyframe diamond.
8. Drag the diamond keyframe up in the graph to **70** (see Figure 10.6).
9. Move forward one more second and then **Option-**click on the line to add another keyframe to the Timeline.

FIGURE 10.5

Video Animations Keyframe Graph.

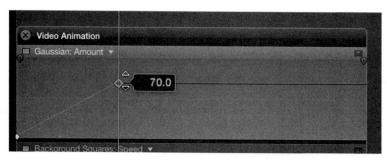

FIGURE 10.6

Keyframe Animation.

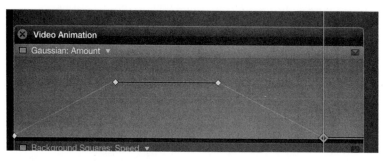

FIGURE 10.7

Complete Keyframe Animation Graph.

FIGURE 10.8

Reset Buttons.

10. Move forward one more second, **Option-**click a fourth time to add another keyframe, and drag the keyframe down to **0** (zero) to complete the animation. The graph should look like Figure 10.7.

11. Close the Video Animation pop-up with **Control-V**.

Play your animations to see how they look. Many effects have parameters like these that can be animated. If you ever get into trouble with your effects, it's often simpler to reset them with the hooked arrow in the Inspector. There's one for every effect, and there's an overall reset button for all the effects (see Figure 10.8).

Audition Effects

We looked at the Audition function earlier in the lesson on editing. Auditioning is most commonly used with the Replace edit function, but it is also a great tool for trying out different effects as well as different settings of the same effect. Let's apply an effect and make some variations to try out:

1. Select the fourth clip in the project, the first use of *posing with doors* and in the Effects Browser go to the Basics group.
2. Find **Colorize** and double-click it to apply it. The effect in its default settings maps white to bright red.
3. Duplicate the clip as an Audition, either by using **Clip>Audition>Duplicate as Audition** or **Control**-clicking and selecting **Duplicate as Audition** or pressing **Option-Y**.
4. The Inspector will now display the duplicate. Use the **Remap Black** control to change the color from a deep red to a deep blue, creating a duotone effect.
5. Click open the **Audition HUD** with the badge in the upper-left corner.
6. Click the **Duplicate** button to duplicate the copy. This time, change the white to bright orange, giving you three available options in the Audition HUD to choose from, as in Figure 10.9.

FIGURE 10.9

Effects Audition HUD.

This is a very powerful tool for trying out different effects in the Timeline and storing them for possible use.

VIDEO EFFECTS

Because there are so many effects in FCP, it would take a whole book to cover them, and there are more being produced every day. There are effects from third-party manufacturers like GenArts and Digital Heaven and Noise Industries as well as many free effects. Like transitions, the effects are all based on Motion 5, and many are being produced just for fun and being shared with others by Alex Gollner, Brendan Gibbons, and others. For a comprehensive collection of free effects go to http://www.fcp.co.

Let's look through the list of effects and pick out some of the highlights, at least the ones I think are useful that you should know about. I suggest spending time just playing with them to see what they can do.

Basics

There are 11 Basics effects starting with the simple **Black & White**, which we skimmed first. It's the easiest way to make video black and white. **Broadcast Safe** is usually applied to whole projects to remove any excess luminance in the images; here's how:

1. Select all the clips in the Timeline and press **Option-G** to make them one long Compound Clip.
2. Select the Compound Clip and double-click the **Broadcast Safe** effect to apply it to all the clips.

This is the equivalent of nesting a sequence in previous versions of FCP. The effect should always be applied last after other color correction and effects have been applied.

There are three tinting effects in Basics: **Sepia**, which allows only sepia tones to be applied to a clip; **Tint**, which lets you change the overall color of the clip; and **Colorize**, which allows you to remap the black to one color and the white to another, effectively creating a duotone. Colorize is really powerful, and you should spend a little time trying it out.

While **Negative** will completely invert the image's luminance values and alter the color as well, the remaining effects in the groups are really color looks. Not sure why they're here, and there are a great many more color looks in the color correction tools, which we'll see in the next lesson.

Blur

There are five Blur effects in the new FCP and one **Sharpen**. You should be careful with Sharpen in video because it creates high-contrast edges that can be very difficult to compress and can easily turn into pixelated edges.

The most commonly used blur is **Gaussian** (pronounced *gowsian*), named after the nineteenth-century German mathematician Carl Friedrich Gauss. In its default settings the effect produces a smooth blurring of the image. It also allows you to blur the images horizontally and vertically separately through a pop-up menu. This capability can be very useful to reducing flickering in an image caused by fine lines and patterning. Here's how to use it:

1. Set the **Horizontal** blur to 0 (zero) while leaving the **Vertical** blur at **100**.
2. Set the **Blur** amount to a very low number like **1.0** or even **0.5**.

The **Directional** blur is one of the many effects that have on-screen controls in the Viewer (see Figure 10.10). Dragging the arrow around the circle will change the direction of the blur, and pulling the arrow farther from the center will increase the blur amount. The wonderful thing about these types of controls, which we will see more of later, is that you can make adjustments right in the Viewer while watching the effect on your video. **Prism**, **Zoom**, and **Radial** also have these on-screen controls.

Distortion

The Distortion effects are, for the most part, more gimmicky than useful, though some, such as **Crop & Feather**, can be practical. The effect gives you controls (see Figure 10.11) for cropping the sides of the image as well as controlling the Roundness, Feather, and the X and Y positions of the cropped area. Many of the effects have values that greatly exceed those shown in the slider. Double-click in the value box for Roundness, for instance, and use the scroll wheel to increase the value. Though the slider control goes only to 50, the scroll wheel will take the value to 1000. This is

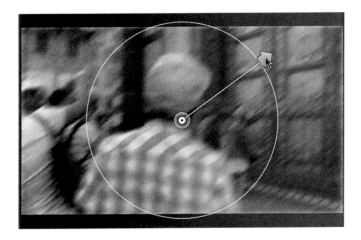

FIGURE 10.10

Directional Blur Controls.

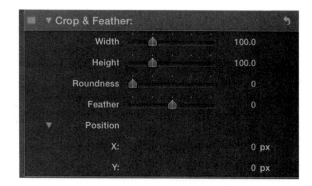

FIGURE 10.11

Crop & Feather Controls.

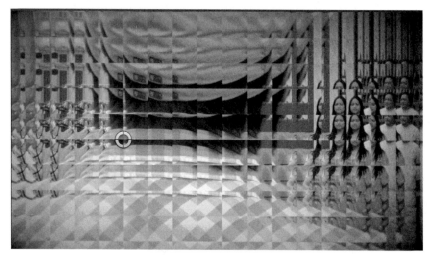

FIGURE 10.12

Glass Blocks.

often the case with many of the value boxes. If the maximum limit of the slider doesn't seem enough, try increasing the value further by scrolling or simply typing a higher value.

Droplet has on-screen controls for Radius and Thicken as well as the position of the effect. The Intensity control is available only in the Inspector, but all of the properties are keyframeable. A nice feature of the on-screen controls is that the keyframes can be added in the Inspector and then adjusted and played within the Viewer.

The real distortion effects are those such as **Background Squares** (which we've seen), **Tinted Squares**, **Fisheye**, **Fun House**, **Insect Eye**, **Mirror**, **Scrape**, **Underwater**, **Wave**, and **Glass Blocks** (see Figure 10.12). These fun effects distort the clip and are often seen in extreme sports videos.

FIGURE 10.13

Flipped.

Earthquake will shake the clip, vibrating it very quickly. If you need that effect, there you have it.

Some effects hidden in the Distortion group are of actual, practical value. **Flipped** is most commonly used in its default settings, where it flips an image from left to right. This capability is really useful if something has been shot in the wrong direction; you might use it if someone should be looking from right to left and instead was shot looking left to right. Flipped will fix that problem. Flipped can not only reverse the image, but also flip it vertically, and both vertically and horizontally; plus, it can control the amount of flipping, a value that can be animated to rotate the image on-screen (see Figure 10.13).

Water Pane produces a very realistic "water droplets running down glass" effect that also gives the shot a bluish cast to emulate a rainy day.

TIP

Nudging Values

If you have a value like the Roundness amount selected in a box and use the **Up** and **Down** arrow keys, you can raise and lower the value in one-unit increments. This is an excellent way to fine-tune your settings and control them with great precision.

Keying

Keying is used to selectively cut out areas of an image. The most efficient way to do this is chromakeying, the technique of removing one specific color from an image. (You see this when meteorologists stand in front of weather maps.) The two most commonly used colors are blue and green. Generally, green rather than blue is usually used for chromakeying video; blue is sometimes used in film production. On the other hand, if your subject has to wear green for St. Patrick's Day, you'll have to use blue.

The key to keying is to shoot it well. Poorly shot material just will not key properly. For chromakeying, the background blue or green screen must be evenly lit and correctly exposed so that the color is as pure as possible. Video has many limitations of color depth and saturation that make good keying difficult. An example of a poorly lit and difficult-to-key shot can be found in the *Effects* event called *green screen*. Folds in the green fabric make the lighting uneven, yet FCP's **Keyer** effect can handle it quite well. Here's how to use it:

1. In the *Effects* event, find the *duomo* shot and append it into the Timeline.
2. Select the shot in the Timeline and press the **X** key to mark a range selection in the Timeline.
3. In the *Effects* event, find the shot called *green screen* and use **Shift-Q** to backtime the shot and connect it to the primary storyline.
4. Select the *green screen* shot and skim the Keyer effect. It doesn't work very well.
5. Double-click the Keyer effect to apply it.

That's much better, though it's still not great. Play this at full screen (**Shift-Command-F**). You can see the crease in the fabric, and as I stand up and walk out of the shot, a shadowing effect falls on the background.

To see this effect better, click the three little buttons next to the View control that changes from Composite to Matte to Original. The Matte view (see Figure 10.14) gives a black-and-white representation of the matte, what's being keyed out displayed as luminance values. What's white is opaque, and what's black is transparent. You can clearly see the shadow on the right and the crease in the fabric.

1. While still in Matte view next to the Refine Key controls, click the **Sample Color** button, the button with the box that has the red dots in the corners.
2. Drag a selection with the tool in the lower right of the Matte image where the shadow image is.
3. In the Timeline, move the playhead forward till I'm standing up and moving out of the shot and you see the highlight area in the upper left.
4. Using the **Sample Color** tool again, draw a rectangle through the highlight; this adjusts the key further.
5. Click back on the first button in the View group to see the Composite image.

FIGURE 10.14

Keyer Matte View and Controls.

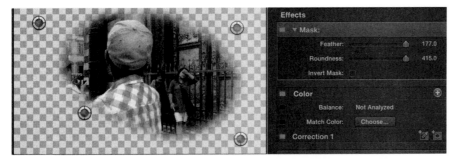

FIGURE 10.15

Matte Controls and Effects.

That's quite a bit better. If you're having trouble, use the **Reset** button to reset the effect and try again.

Other useful controls are the Strength slider, which defaults to 100%, but using the scroll wheel on your mouse, you can push it up to 200%. Fill Holes is handy if you have small portions of the keying color isolated in the image. Edge Distance fine-tunes the image when there are wispy edges like hair that are being cut off. Spill Level is important if the green from the background is spilling onto the subject. It adds magenta into the image to help counteract that problem.

The **Luma Keyer** effect works best on images with a bright white background. It's much easier to key white than black because almost every image will have some degree of shadow area in it, which keys out. The Luma Keyer can be useful to create ghostly overlay effects.

The **Mask** effect is far more useful than its simple controls would lead you to believe (see Figure 10.15). The on-screen target buttons position the corners of the four points of the matte shape. Feather softens the edges, and Roundness smoothes out the shape. Notice in the graphic that these two controls go to much higher values than the slider and allow you to create organic shapes and effects that let you layer multiple images over a background effectively. This is an excellent four-point "garbage matte" tool.

Light

There were only a few Glow effects in the last few versions of FCP, but there are 14 Light effects in this group. You can try them out for yourself with FCP's preview functions. I suggest you not only skim the effects, but also use the play feature by selecting the effect and pressing play. Some of the effects, such as **Intro Flashes** and **Quick Flash/Spin**, which are similar, affect just the beginning of the image, almost like a transition, zooming into them, twisting and flashing the picture at the start.

Many of the effects have interesting controls that allow you to alter the position and color of the light effects. Some of the effects, such as **Artifacts**, **Highlights**, the

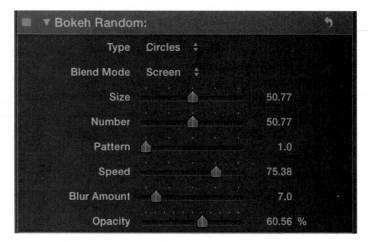

FIGURE 10.16

Bokeh Random Controls.

Flash effects, **Side Lights**, **Subway Shadow**, and **Bokeh Random**, have motion built into the them by the movement of light. The others, such as **Aura**, **Bloom**, **Dazzle**, **Shadows**, **Spot**, and **Glow**, are static and take their effect from the highlights in the image.

Some effects, such as the **Flash** effects, have little or no controls at all, whereas others, like **Bokeh Random,** have extensive controls (see Figure 10.16). These controls include changing the bokeh type as well as any of the extensive composite modes in FCP. You can also control the size and number and pattern and speed and blur and opacity of these bokeh objects, resulting in effects from authentic-looking lens flares to patterns resembling Japanese lanterns floating through the frame. It really is a lot of fun to play with the effects they produce, particularly on telephoto shots that often produce the effects naturally.

Spot can be useful to highlight someone or something on the screen (see Figure 10.17, also in the color section). It's much more subtle than just a circle, yet when you increase the contrast and control the radius and the amount of feather in the image, this effect quickly leads the viewer to the point of interest. The target position and the other controls can be easily animated if the subject moves around the screen.

NOTE

Bokeh

Bokeh is a Japanese word that means blur or haze; it refers to very out-of-focus highlights in the frame that appear as shapes of light with no definition. The shape of the bokeh, usually circular or hexagonal, can be caused by aberrations in a camera lens.

FIGURE 10.17

Spot Image.

Looks

Most of the Looks are really color effects, and most have very few controls except for an Amount slider. Many of them also have an important slider to protect skin tones.

The **50s TV** effect is basically low-contrast black and white, and **Bleach Bypass** is high-contrast desaturated video made popular in the movie *Saving Private Ryan*. **Combat**, **Strife**, and **Numerik** are similar, with more controls as well as skin protection.

Cold Steel and **Cast** have Protect Skin sliders, which you need to push up to take effect. **Cast** allows you to change the color shading, so it's almost a tint effect with some skin tone protection. **Dry Heat**, **Faded Sun**, and **Indie Red** also have skin protection to prevent the flesh tones from going too brown. The slider adds more blue back into flesh colors. This is true also of **Teal & Orange**, which is an essential effect for every Michael Bay movie or if you're trying to emulate the *Transformers* look.

Some of the warmer looks such as **Desert Glare** and **Heat Wave** don't have a skin protection slider.

Cool Tones and **Day into Night** are very similar: very cold and dark. The first has more controls with less overall blue.

Glory, **Romantic**, and **Memory** are a cross between a color look and a light effect. The first gives a light rays effect that has an on-screen control that you can position where you like and animate if you wish.

Isolate is different, with a pop-up that allows you to desaturate either skin tones or shirts. With shirts selected, it removes all color except for skin tones. The default, which is to desaturate skin and make it grayscale, I find rather ugly.

Night Vision is a unique look. It's a deep green color effect together with a matte, binoculars, a circle, an ellipse, or no matte at all. There's a Size slider, to adjust the size of the matte; a Grain slider, which I think helps the effect; and a Light slider, which controls the brightness. The Amount dims the matte, so I don't think it works very well with the effect.

Stylize

The Stylize group is a real grab bag of effects. Some are very stylistic, whereas others are very practical, many of them based on film-style effects such as the **Super 8mm** effect.

Add Noise lets you add different types of grain—from horizontal TV graining to film grain in different colors. It works well when combined with Looks. **Bad TV** has some of the same grain effects, plus scan lines, and horizontal shake. It's really pretty bad TV. **Raster** is similar without the wobbly image. **Add Noise** also works well when used after **Aged Film**, which gives your video a sepia tone and adds film scratches. **Projector** gives film flicker, but without the sepia tone. **Aged Paper** takes sepia to an extreme by adding a paper mask over the image. The one slider makes the mask size bigger and smaller. At zero, you see only the paper and no image, and at 100, you see the whole image blended with the aged paper. **Film Grain** and **Cross Hatch** are also sepia-tinted effects. The latter combines with a criss-cross line pattern. Be careful with this one because it is very difficult to compress well and will pixelate easily. Be careful also with **Graphic**, **Halftone**, **Sketch**, and **Line Screen**.

Camcorder gives you a camera viewfinder look, with a blinking red record light.

Frame is a stylized border effect that reduces the size of the image and adds a choice of one of a dozen different border looks. These aren't simple colored edges, but actual picture frames, as in Figure 10.18. For a simple color border, use **Simple Border**. **Photo Recall** is a similar border effect, only not over transparency, but over a blurred version of the image.

Three of the most commonly used style effects I've saved for last. **Letterbox** lets you take a standard 4:3 video and crop it to one of five standard cinema shapes to create a letterbox effect. This is a crop, not an overlay, so the area outside the image is empty. If you want to place a color there, you should put one of the color mattes underneath it.

If you want to make a whole project widescreen—which is probably the point rather than applying it to individual clips—you need to convert the project to a Compound Clip and apply the effect to the Compound Clip like this:

1. Select all the clips in the Timeline and use the **File>New Compound Clip** or press **Option-G**.

FIGURE 10.18

Frame and Controls.

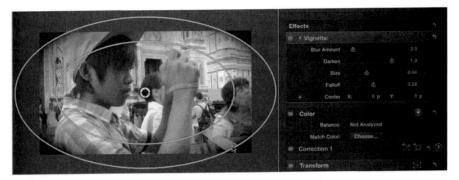

FIGURE 10.19

Vignette and Controls.

2. Select the Compound Clip in the Timeline and apply the Letterbox effect to it.
3. Change the Aspect Ratio effect controls in the Video tab of the Inspector.

Vignette has complex on-screen controls (see Figure 10.19) that allow you to darken the edges of the image, control how much dark appears, and determine how rapidly it falls off. You can also position the Vignette so it does not have to be centered. It's a very popular stylistic feature that's used to some degree in a great many shots. You can also do this in the color controls, as we'll see in the next lesson.

Censor lets you blur or pixelate a portion of the screen. It can be someone's face, a T-shirt, a license plate, anything. Here's how to use it:

1. In the Timeline, select the first *posing with doors* clip and delete any effects that have been applied to it.
2. With the Timeline clip selected, double-click the Censor effect in the Effects Browser to apply it.
3. Make sure the clip is selected in the Timeline, which makes the on-screen controls visible in the Viewer.
4. Drag the center target to position the pixelization over the boy's face.
5. In the Inspector controls, you can change it to **Blur** or **Darken**.

Tiling

Tiling is a category of four effects and includes **Kaleidoscope** and **Kaleidotile**, which can be animated to produce kaleidoscopic effects. They work especially well on images with motion in them because they rely on movement in the frame to create movement in the effect. With a still image, the kaleido effect will simply be stuck on the image unless you animate the properties.

The **Tile** effect is the Replicate effect of previous versions and can be changed from one, which is no effect, up to 20, which is 20 rows of very small images (see Figure 10.20).

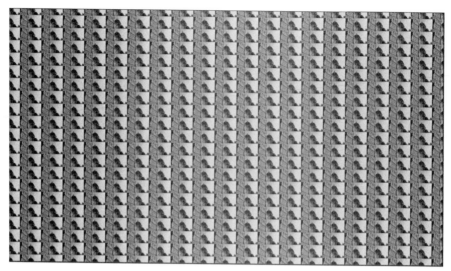

FIGURE 10.20

Tile Effect.

IMPORTANT NOTE

Deinterlacing

An important feature is the ability to deinterlace video, especially when adding interlaced video such as DV or 1080i60 media to a progressive project. There is no Deinterlace effect in FCP as in previous versions. However, you can deinterlace clips either in the Timeline or in the browser before they're added to the project. Select the clip or multiple clips and go to the **Info** tab of the Inspector. In the lower left, from the pop-up that normally is **Basic View**, select **Settings View**. In the **Field Dominance Override** pop-up, change that from whatever field dominance is shown to **Progressive**. This will effectively deinterlace your media. To see the effect on your video, make sure you have **Show Both Fields** switched on with the display controls in the upper right of the Viewer.

AUDIO EFFECTS

There are even more audio effects than there are video effects, many the province of the serious audiophile, which I confess I am not. There are so many effects in part because of some redundancy between FCP effects; Logic effects, which have been ported to this application; and Mac OS X effects, which also appear.

There is a short project in the *FCP6* disk that we will use to apply some effects:

1. In the Project Library select the *Audio* project in the *FCP6* disk and use **Command-D**.
2. Duplicate just the project to your system drive or your media drive, not to the *FCP6* disk.

3. Do not duplicate the associated events or render files.
4. It's a good idea to rename the duplicate something like *Audio copy*.

You apply an audio effect just like you do a video effect. You can preview it by playing or skimming the effect.

The **Distortion** group of effects includes such popular favorites as **Car Radio** and **Telephone**. You use it like this:

1. Select the last clip in the *Audio copy* project.
2. In the Distortion group of audio effects, find **Telephone**.
3. Press play to preview the effect and hear the way the voice has been reduced to the narrow telephone frequency range.
4. To apply the effect, double-click it.
5. To access the Telephone controls, go to the Audio tab of the Inspector (see Figure 10.21).
6. Try different telephone types from the preset pop-up.
7. Click the Channel EQ and Compressor icons that bring up HUDs showing you what's being done to the audio and giving you even further control of the effect.
8. The Amount slider will adjust how strong the Telephone effect is. As you adjust the slider, notice how it alters the EQ and Compressor controls.
9. With the clip selected in the Timeline, press the **forward slash** (/) key to play just the selection.

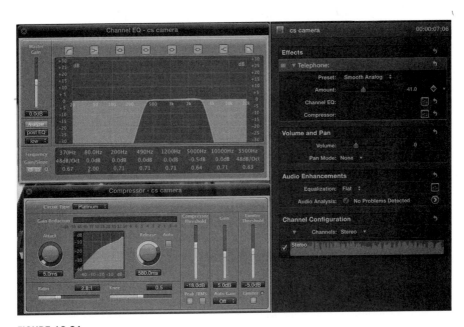

FIGURE 10.21

Telephone Controls.

> **NOTE**
>
> Looped Selection Playback
>
> When you make adjustments while playing a selection using the **forward slash** key, playback may fail to loop back at the end of the selection. I hope that this issue will be rectified by the time you're reading this.

There are numerous powerful and varied distortion effects. From the Logic subgroup, try the **Clip Distortion** effect, which itself has a great many distortion presets. In fact, all of the Logic effects have many presets. It will take you hours to try them all out. **Ringshifter** alone has three submenus of presets with a total of 30 effects. Have fun!

> **TIP**
>
> Soloing
>
> If you have multiple layers of audio, it's useful to "solo" the audio. You can do this by clicking the **Solo** button next to Snapping in the upper right of the Timeline or by pressing **Option-S** for soloing.

The **Echo** group has nine effects, all with a subtle difference in the tonal quality they apply. Many of them, such as the Logic effects **Delay Designer,** have many more presets for you to try and to adjust further with the Delay Designer HUD.

1. In the *audio* Keyword Collection, find the jazz piece called *music*.
2. Append it to the project with the **E** key.
3. Select the music in the Timeline and press the **forward slash** key to play a little of it.
4. Stop playback and select one of the Logic effects like Delay Designer or Stereo Delay.
5. Press the **spacebar** to hear the effect.
6. Double-click the effect to apply it and use the **forward slash** key to play the music while you try different presets.

Most of the Final Cut EQ effects are basically handy presets for one of the available equalizers, usually **Fat EQ**, which gives the most range of controls.

The Logic subgroup of EQs simply gives you direct access to **AutoFilter, Channel EQ, Fat EQ**, and **Linear Phase EQ,** all set flat for you to adjust as desired. Personally, I like the frequency dragging controls of the Channel EQ HUD the best (see Figure 10.22).

In the **Levels** group of effects, the Logic **Adaptive Limiter** at its default settings applies a little compression to the audio that's very useful. It has the overall effect of providing a little gain without increasing background noise. Because of this boost in the gain, you might need to pull down the level a few decibels. On music such as the jazz piece in the project, try the **Adaptive Limiter** with the **Add Density** preset.

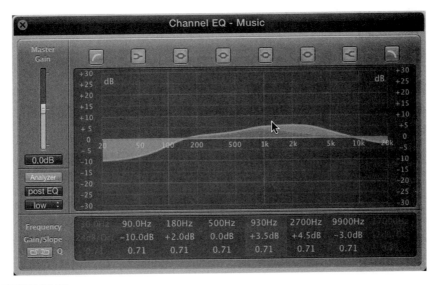

Channel EQ - Music

Frequency	90.0Hz	180Hz	500Hz	930Hz	2700Hz	9900Hz
Gain/Slope	−10.0dB	+2.0dB	0.0dB	+3.5dB	+4.5dB	−3.0dB
Q	0.71	0.71	0.71	0.71	0.71	0.71

FIGURE 10.22

Channel EQ HUD.

The **Modulation** group has mostly special effects for music such as **Tremolo, Flanger, Chorus**, and **Vibrato**. The Logic Tremolo is pretty extreme even in its default settings. You probably don't want to use any of these on voice, but they make some interesting sounds for music.

Spaces creates different types of room reverberation.

1. Select the *looking up* clip in the project and press **backslash** to play it.
2. Stop playback of the clip, and in the Spaces group of audio effects, click on **Cathedral**.
3. Press the **spacebar** to play the clip with the effect.
4. Try the **AUMatrixReverb**.

The best tool for creating room acoustics is Logic's **Space Designer**. In addition to having its own design HUD, it has a staggering number of presets (see Figure 10.23) for all types of large, medium, and small spaces, with different types of rooms and halls, from Michaelis Nave to Villa Bathroom, and everything in between.

Though there are no bars and tone available in FCP, any clip with audio can be used with the **Test Oscillator** in the **Specialized** group of audio effects:

1. Select the *music* clip in the project, and in the Specialized group, find the **Logic Test Oscillator**.
2. Double-click the effect to apply it and play the clip in the Timeline. You hear the default 1000Hz tone at −12dB.
3. In the Inspector, you'll find a whole list of different presets (see Figure 10.24), including sweep tones and a HUD to set your own tone presets.

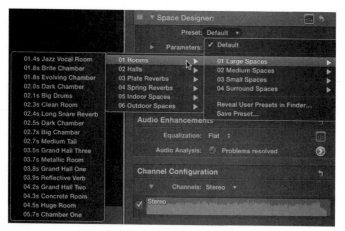

FIGURE 10.23

Space Designer Presets.

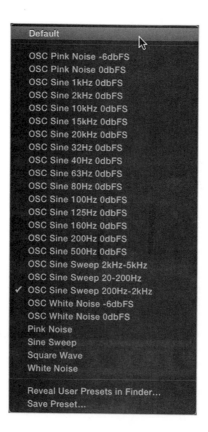

FIGURE 10.24

Test Oscillator Presets.

Strangely, in the **Voice** group, most of the Final Cut effects are comic effects, like **Alien**, **Cartoon Animals**, **Robot**, and **Helium**. In special cases, you might find **Disguised** useful.

One useful effect in the Logic subgroup is **DeEsser** for voices that are too sibilant. **Vocal Transformer** has some interesting presets as well.

There's a lot to try, so take your time and play with the different effects. Try them with your own recordings and see what you can do to enhance them.

RETIMING

Retiming, which was called Speed in earlier versions of FCP, has been completely changed in the current iteration of the application. These aren't really effects in the true sense, but they're used as special effects, so I've included FCP's Retiming functions in this lesson.

Constant Speed Changes

Retiming can be accessed from either of two menus or in the Timeline itself. You can access Retiming by

- Using the Retime menu in the Toolbar (see Figure 10.25)
- Using Retime in the **Modify** menu
- **Control**-clicking on a clip and selecting **Retime**
- Selecting a clip in the Timeline and pressing **Command-R**

FIGURE 10.25

Retiming Menu.

To work with the Retiming tools, we're going to use some of the winter clips from the *FCP5* disk image:

1. Start by making a new project called *Retime* on your system or media drive.
2. In the *Snow* event, find the *trash can1* shot.
3. Drag a selection from around 6:21, just before the snowboarder enters the frame, to 8:17, just after he disappears.
4. Append the shot into the Timeline.
5. Make a selection in *trash can2* starting at about 3:13 and ending about 5:12.
6. Append that shot into the Timeline.
7. From *trash can3* make the selection from 29:28 to 32:20 and append that shot.
8. Select all the clips in the Timeline, and from the Retime menu in the Toolbar, select **Slow 50%**.

All the shots will slow down to half speed and will appear in the Timeline showing the orange retiming bar. You're not confined to the slow and fast speeds that appear in the Toolbar, however.

To make the last shot in the project a bit faster, grab the drag handle at the end of the Retiming bar and pull it to the left to speed up the clip, as in Figure 10.26. The speed is now 60%, as indicated on the Retiming bar.

To make the clip play backward, select it, and from the Retime menu, select **Reverse Clip**. The Retiming bar in the Timeline will have backward-arrow marks on it and the speed will be displayed as −60% in this case (see Figure 10.27).

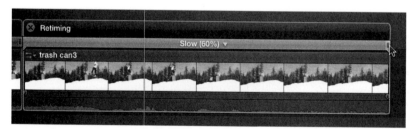

FIGURE 10.26

Changing Clip Speed.

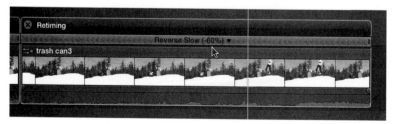

FIGURE 10.27

Reverse Clip.

1. Select the middle clip in the project and use the Retime menu to reset its speed or press **Option-Command-R**.
2. Click on the disclosure triangle next to the speed in the Retiming bar and select **Fast>2x**.

The clip will be shortened, the Retiming bar will appear in blue, and the speed will be indicated as 200%, twice normal speed. There are other special retiming effects you can apply:

1. Select the second clip in the project, and from the Retime menu, select **Instant Replay**. The clip plays twice.
2. Select the first clip, and from the Retime menu, choose **Rewind>2x**. The clip plays, rewinds at twice normal speed, and then plays again.
3. Select the first segment of the clip that's at half speed and click the disclosure triangle to choose **Normal (100%)**, or click on the orange bar and press the shortcut **Shift-N** to reset that segment of the clip. Now the clip plays forward at normal speed, rewinds at double speed, and then plays forward in slow motion.

Not only does the Rewind clip play backward, but if you zoom into the Timeline and look closely, you'll see that the Instant Replay effect has one frame where the video is resetting backward.

To reset all the clips back to normal so that we can look at other features, select them all and use **Option-Command-R**. Then press **Command-R** to close the Retiming bars in the Timeline.

NOTE

Phase Shifting

You may have noticed that, at the bottom of the Retime menu, **Preserve Pitch** is checked on by default, which is probably what you want. Though the clips may change speed, the audio is automatically adjusted and phase shifted so it does not sound like Mickey Mouse when it's speeded up, nor does it become deep and drawn out when slowed down. In fact, voices are quite understandable when speeded up or slowed down.

Variable Speed

Speed changes thus far have all been constant speed changes, even when rewinding and replaying, but sometimes you want to speed up a clip or slow it down over time. The two simplest ways are to use the Speed Ramps:

1. Select the first clip, and from the Retime menu in the Toolbar, select **Speed Ramp>from 0%**. The clip is broken down into segments, each getting progressively quicker over the course of the clip (see Figure 10.28).
2. Select the second clip and put the playhead right where the skier's hand touches the trash can. Then, from the Retime menu, select **Hold** or press **Shift-H**. A hold frame is created that's two seconds long. The clip plays in normal speed, stops, and then plays forward.

FIGURE 10.28

Speed Ramp.

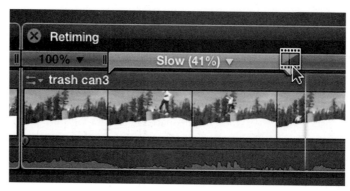

FIGURE 10.29

Changing End Frame.

3. Drag the handles on the ends of the red retiming bar for the hold segment to adjust the duration of the freeze.
4. Hold the **R** key and drag a range selection in the third clip beginning just as the snowboarder leaves the snow until he's cleared the trash can, about 24 frames. When you release the **R,** the pointer will return to the Select tool.
5. From the Retime menu, select **Slow>50%**. The middle segment is slowed down.
6. Drag out the ends of the orange retiming bar to make the speed of the middle segment even slower.
7. Once you have the speed as you like it, click the disclosure triangle on the middle segment and select **Change End Source Frame**.
8. A frame icon appears at the end of the segment (see Figure 10.29). Drag it to adjust the end frame. The speed of the segment doesn't change, but the last frame before speed resumes is altered.
9. Once you've adjusted the end frame of the middle segment, the same option appears in the first segment and lets you adjust the frame for the end of that segment.

Whenever you speed up or slow down a clip, the length of the project changes; it ripples as the clips get longer or shorter. But sometimes you don't want the project to change duration. It's fairly simple to do. Let's start by resetting the middle clip:

1. Select the middle clip and press **Option-Command-R** to reset it.
2. Use **Option-Command-Up** arrow to lift it out of the storyline as a clip connected to a gap.

3. Change the speed of the clip to 50% or whatever you like. The Connected Clip extends over the third clip in the project.
4. Trim the end of the shot so it's the same length as the gap.
5. Activate the Trim tool (**T**) and slip the contents so the skier is in the frame at the beginning of the jump and completes it by the end.
6. Select the clip with the Select tool and press **Shift-Command-Down** arrow to overwrite the Connected Clip back into the primary storyline.

There is another way to do this that is a little more complex but does the job in fewer steps:

1. Select the clip you want to speed change, either speed up or slow down, and press **Option-G** to make it a Compound Clip.
2. Double-click the Compound Clip to open it into the Timeline. This Timeline is constrained to the duration of the clip in the main project, but inside the Compound Clip you can do anything you want to it.
3. Change the speed of the clip to 50% or whatever you like.
4. Use the **Trim** tool with the **Slip** function to adjust the contents, or if you've speeded up the clip, stretch it out to make it longer.
5. Switch back to the project with the **back** button to see how the speed change looks in the project.

You may have noticed a couple of other items in the Retime menu. One is the **Conform Speed** function. Normally, when clips that are at high frame rates, such as 50fps, are placed in 25fps projects, the clips run at 25fps. One second of 50fps media is still one second in the 25fps Timeline. However, if you use Conform Speed on the 50fps clip, it will now run half a second for every second in the Timeline, effectively slowing the clip down by 50%. This is the smoothest way to get good slow motion. You're actually using the frames shot in the camera to produce the slow motion effect. This is much better than using frame interpolation by the application to create the effect.

The other option you'll see in the Retime menu is the **Video Quality** submenu. You have three quality options: **Normal**, **Frame Blending**, and **Optical Flow**. Normal simply duplicates or discards frames as needed. Frame Blending creates intermediate whole frames as needed and generally should be used for slow motion effects. Optical Flow will produce the best results. It's based on analysis of the image content, calculating frames based on what's moving in the frame to create the smoothest motion. This should be used for extreme slow motion effects. This can be very slow to render, however.

STABILIZATION

Stabilization analysis can be done when clips are imported, but unless you have a lot of material that needs to be analyzed, then it's probably better to do it after the material is in your system. You can do the analysis to either the clip in the event or in the Timeline. It makes no difference; it's done to the entire clip, and the analysis files are

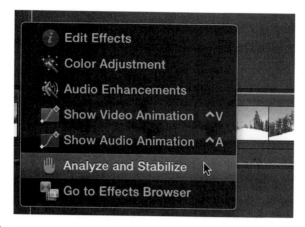

FIGURE 10.30

Analyze and Stabilize.

used by every instance of the clip you work with. As this material was not analyzed when it was imported, we'll do it now:

1. To analyze the clip in the event, you could select the *hand held trash can* clip in the *Snow* event and **Control-**click to choose **Analyze and Fix** or use **Modify>Analyze and Fix**.
2. Before you analyze it, edit the clip into the Timeline by appending it.
3. Click the badge in the upper-left corner of the clip and select **Analyze and Stabilize** (see Figure 10.30).

The nice thing about doing the analysis in the Timeline is that stabilization is automatically turned on. If you do the analysis only in the event, you still have to turn the stabilization on for the clip in the Inspector.

SUMMARY

This chapter covered just the tip of the iceberg of some of the video and audio effects in FCP. I urge you to look through them and explore their capabilities for yourself. In addition to the Effects Browser content, we looked at working with Retiming and with Stabilization. While Color Correction is often thought of as effects, they use an entirely separate utility that's built into every clip, which we'll see in the next lesson. By now, you should have a fairly good idea of what you can do with this application and should be well on your way to creating exciting, interesting, and original video productions.

Color Correction

11

LESSON OUTLINE

Color correction in FCP, unlike in previous versions of Final Cut, is not done using effects, or filters as they were called. Instant color correction is an integral component of every video clip and accessed like so much else in the application—using the clip's Inspector.

For this lesson we'll use *FCP6.sparseimage*, which we used in the preceding lesson. If you haven't downloaded the material, or the project or event has become messed up, download it again from the www.fcpxbook.com website.

1. Once the file is downloaded, copy or move it to the top level of your dedicated media drive.
2. Double-click *FCP6.sparseimage* to mount the disk.
3. Launch the application.

> **NOTE**
>
> H.264
>
> The media files that accompany this book are heavily compressed H.264 files. Playing back this media requires a fast computer. If you have difficulty playing back the media, you should transcode it to proxy media as in previous lessons.

Before we get started, let's make a new project to work with:

1. In the Project Library, select your media drive and use **Command-N** to make a new project.
2. Name the project *Color* and set the *Effects* event as the default event.
3. In the *fix* Keyword Collection in *Effects*, select all the clips and append them into the Timeline. They will go into the project in the same order they are in the Event Browser.

COLOR CORRECTION

Good exposure and color begin in the shooting. It's always easier and better if you shoot correctly from the beginning rather than try to fix these issues in postproduction. That means lighting the scene well, exposing it correctly, setting your white balance correctly, and not leaving the camera's auto exposure and white balance to do the guessing.

The color correction tools in FCP are professional-strength tools, so use them carefully. A new feature in FCP is to rely on Apple's ColorSync to coordinate the color display and keep it consistent throughout the production process, from first importing, through editing, and for final sharing and export.

Color Balance

FCP has automatic color balancing tools that can be activated in three different ways:

- Using the **Modify>Balance Color** selection in the menus
- Clicking the Color Balance button in the Inspector (see Figure 11.1)
- Selecting **Balance Color** from the Enhancements menu in the Toolbar

FIGURE 11.1

Color Balance Checkbox.

When Color Balance is applied, the clip is automatically adjusted based on a mathematical formula. For many users, this will be their one stop to color correction, and it will be good enough. If you are used to working with Sony cameras, the look Color Balance produces is probably acceptable. Sony cameras tend to produce results like these. For me, this feature is a little desaturated, but more importantly, it seems to have a bluish cast to it. Here's how to use it:

1. Move the playhead in the Timeline toward the end of the *silhouette with doors* clip and select it.
2. In the Inspector, find the **Balance** function under Color. If it's not visible, double-click **Color** to open the correction tools.
3. Check on the **Color Balance** button so the LED is blue.

The arm of the woman holding the camera has a decidedly bluish cast to it. There should really be no blue in human flesh tones unless the light has a bluish tone, such as at dusk, either before sunrise or well after sunset. Normally, flesh tones do not exhibit any blue. I also find the color a little weak. That said, the tool is very good at making an instant correction of problem footage, but for most shots, you would be well served to color correct your images manually. Not only will your video look great, but it's great fun to add special color looks and effects to your video.

TIP

Compressed Video and Color Correction

If you're working with proxy video or heavily compressed video formats like AVCHD and MPEG-4, it's a good idea to transcode the media to ProRes, as this will handle heavy processing much better than compressed files.

Video Scopes

To adjust your color, you need to use the tools that FCP provides for assessing the color. These are the Video Scopes, which can be called up from the **Window** menu or with the keyboard shortcut **Command-7**. You can also open the Video Scopes by using the switch in the upper right of the Viewer and selecting **Show Video Scopes**.

The default video scope that opens is the **Histogram** (see Figure 11.2, also in the color section). This shows the percentages of different luminance and colors in the image. This is the Histogram of the *cheese* clip. Pure black is zero and is on the left. Peak white is 100 and is on the right. Notice that this shot, and every shot here, and most every shot taken by most every consumer camera will always record at video levels that exceed peak white and are not broadcast legal. If you are planning to put your video on a television set, every shot will need to be corrected. By the way, the built-in automatic Color Balance does not correct sufficiently for these types of cameras in many cases, and peak levels still exceed television legal standards, which is yet another reason to manually color correct your video. Apart from the video that exceeds 100% white, you can see from the Histogram that this is a fairly even distribution of light to dark across the image. Figure 11.3 is the Histogram of the second

FIGURE 11.2

Histogram.

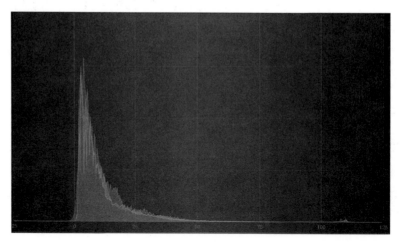

FIGURE 11.3

Histogram of Dark Image.

shot in the project, *ms musician*. As you can see from the image and the Histogram, the vast majority of the luminance values and even of the color values are all to the left, and almost nothing is to the right of 50%.

While the Histogram is the first video scope that appears, it is perhaps the least used in video production. Perhaps the most important scope for video is the **Wave-form Monitor**, which you switch on from the Settings menu at the top left of the scopes or with the shortcut **Shift-Command-7**. Figure 11.4 shows the *silhouette* shot on the right and its waveform on the left. This is also a representation of the luminance

FIGURE 11.4

Waveform Monitor and Image.

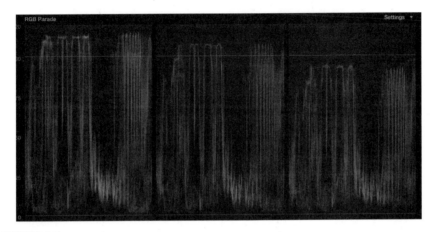

FIGURE 11.5

RGB Parade.

values in the image, but displayed just as it is on the screen, reading across the image from left to right. The bottom of the trace is zero, which is black, while the top is 100. Again, the bright areas in the image exceed 100 units or IRE, which is your target for peak white, while zero is your target for pure black. You can see in the waveform the bright vertical sections in the trace that correspond to the bright areas of the wall behind the woman. The Waveform Monitor is your guide to setting your exposure and your video levels and is a critical tool in the color correction process.

With the Settings Display still set to the Waveform Monitor, under Channels, select **RGB Parade** (see Figure 11.5, also in the color section). This is an important tool for color balancing. It displays three separate views of the image, showing amounts of red, green, and blue in the image. Ideally, you want to have equal proportions of red, green, and blue in the shadow areas of your image, black being the absence of color. If the three colors are balanced in the shadow areas, they are neutralized, and the image will show no color cast in the blacks. This image has a slight elevation in the blue levels at the bottom of the display, slightly higher amounts of

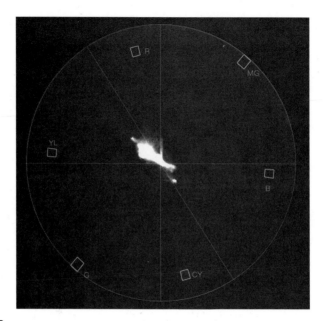

FIGURE 11.6

Vectorscope.

blue than of red and green, probably not something you would notice with your eye, but something the scopes clearly see. On the bright end, you also try to balance the reds, greens, and blues so that equal quantities appear in the highlights, so that white is white and not some shade of a color. Similarly anything that's gray in the image should also show equal proportions of red, green, and blue.

The fourth scope used in video is the **Vectorscope**, which you select in the Settings menu or invoke with **Option-Command-7** (see Figure 11.6, also in the color section). This is not a representation of the image at all, but a display of the amount of which colors are in the image. You can see a faint color wheel ringing the scope and the target boxes for pure colors that would be hit by an SMPTE color bar chart, starting with red, just to the right of 12 o'clock, and going clockwise to magenta, blue, cyan, green, and yellow. Notice the diagonal line that goes from the upper left to the lower right through the circle of the scope. This is the flesh line. Any human skin in the image, if lit by neutral light, correctly balanced, will fall on this line. If it doesn't, you have a color imbalance.

These are the scopes that we'll be switching between as we correct our color.

Color Board

The primary tool for color correction in FCP is the **Color Board**, which is accessed with the arrow button opposite Correction 1 in the Inspector (see Figure 11.7) or more simply with the keyboard shortcut **Command-6** (see Figure 11.8, also in the

FIGURE 11.7

Arrow Button to the Color Board.

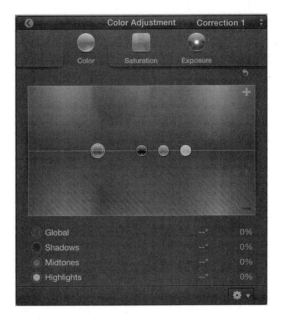

FIGURE 11.8

Color Board.

color section). You can close the Color Board with the same shortcut or with the arrow button in the upper left. You can also access the Color Board from the **Window** menu, from the Enhancements menu in the Toolbar, and from each clip's badge where it's called **Color Adjustment**. Five ways should be enough.

The Color Board has three tabs. The first that we see when we open the tool is the **Color** tab. The second is the **Saturation** tab, and the third is the **Exposure** tab (see Figure 11.9). The last two look almost identical.

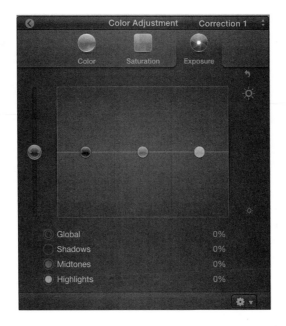

FIGURE 11.9

Exposure Tab.

The Exposure tab has four buttons or pucks, which can only move vertically, up and down. On the left, the larger puck controls the overall exposure of the image, raising and lowering the brightness. Inside the Exposure board are three separate pucks to raise or lower the Shadows, Midtones, and Highlights. These are not sharply defined regions but ones that have considerable overlap.

Saturation is the amount of color in an image, whether it's colorful or washed out, or even black and white. The Saturation tab looks the same as the Exposure tab, with four pucks. The large one on the left controls the overall saturation, and the three others control how much color there is in Shadows, Midtones, and Highlights.

The Color tab also has four controls. The larger puck raises and lowers the overall color in whatever hue you want. Here, the pucks move not only vertically, but also horizontally. If you want to add more blue to the overall look of an image, you take the large puck, move it to the blue section of the Color Board and push it upward. Up will add more of a certain color, and down will remove a certain color. There are three smaller pucks that let you change the color values in specific luminance ranges: Shadows, Midtones, and Highlights.

Using the Color Board

Though the Color Board is laid out from left to right—Color, Saturation, Exposure— you should always make your color correction in reverse order. Always set the Exposure first using the Waveform Monitor displaying Luma, then set the Saturation using the Vectorscope, and finally set the Color using the Waveform displaying the RGB Parade. Though you have these great scopes in FCP to guide you, your primary tools for color correction are your eyes and, ideally, a properly calibrated display. If you watch television or go to the movies, you know that color in video and film can be anything you want it to be. The only goal you should try to maintain is to make sure your images are within specification. Fortunately, you have an effect, the Broadcast Safe effect, that makes this easy for you.

That said, let's start our color correction with one of the images in the project, the *cheese* shot. The order in which your browser is set may determine the way the project displays. In my project the *cheese* shot is the first shot. Proceed like this:

1. Select the *cheese* shot, open the Inspector, and opposite Correction 1, click on the arrow button for the Color Board.
2. Open your **Video Scopes** and switch to the **Waveform Monitor**.
3. In the Color Board, go to the **Exposure** tab.
4. There's nothing really wrong with the overall luminance levels of this shot. Pull down the **Shadows** puck to bring the blacks down to zero.
5. Pull down the **Highlights** a little to keep the whites under 100 IRE.
6. Push up the **Midtones** to brighten the look of the image and add a little pop to it. After you adjust the mids, you'll probably have to adjust the **Shadows** and **Highlights** again.

The process of color correction is a constant ping-ponging of pushing and pulling, raising one value and pulling down another to offset it in some areas.

Let's switch to the Saturation tab. The saturation for the image is pretty good. I suggest just pushing up the overall chrominance level a little to give it a little more vibrancy.

And now we come to Color. Again the color isn't bad, except for a slight blue cast in the highlights and a little too red in the shadows. To correct this:

1. Switch the Color Board to the **Color** tab and switch your scopes to the **Waveform Monitor**.
2. In Settings, under Display, select RGB Parade.
3. Take the **Shadows** puck over the right side in the red and pull the red down a bit.
4. Take the **Highlights** puck and put it to blue and pull it down a little.
5. Finally, to warm the image a little, take the **Midtones** puck over to yellow and push it up a little so the Color Board looks something like Figure 11.10.

TIP

Nudging the Pucks

The pucks can be moved around the Color Board with the arrow keys. Select a puck and use the **Left** or **Right** arrow keys to move it horizontally, and the **Up** or **Down** arrow keys to move it vertically. This lets you move the puck with much greater precision than you can simply by dragging it, though it probably is not nearly precise enough for a professional colorist.

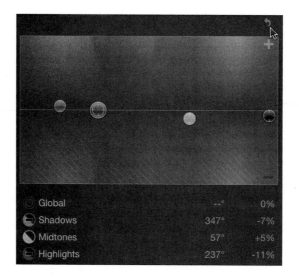

FIGURE 11.10

Corrected Color Board.

Notice at the bottom of the Color Board the display that shows you the color values you're using, also indicating with the icon that shadows are in negative red, mids in yellow, highlights in negative blue, and global is unchanged.

If you want to reset the Color tab, click the hooked arrow button in the upper left. There's a separate reset for Color, Saturation, and Exposure. To see a before-and-after look or to reset the entire Color Board, go back to the Inspector and use the reset button opposite Correction 1. To toggle a before and after, click the blue LED next to Correction 1.

TIP

Reset

To reset a single puck, such as the Shadows puck, click on the puck to select it and press the **Delete** key. The puck will return to the mid line in its default position. This works for any of the pucks in any of the tabs.

White Balance

Let's look at an image with the opposite problem—a white balance with too much yellow. The shot we're going to work on is the last shot in the project, the clip called *statue dutch*. This shows a fairly common problem with consumer cameras. They often read a scene's color as either too warm or too cold. In this instance, the camera made the image overly yellow. I tried the automatic color balance but found it very unsatisfactory. While it corrected the yellow cast, it added a little too much blue and desaturated it so much as to make it almost colorless. Let's see if we can fix this better using the Color Board:

1. As always, we'll start in the Exposure tab of the Color Board. Using the Waveform Monitor displaying Luma as a guide, take down the **Shadows** a unit or two with the **Down** arrow.
2. Take down the **Highlights** so the white sky in the upper left of the scene is touching 100 IRE.
3. Push up the **Midtones** to brighten the shot to taste. I used 10 taps on the **Up** arrow.
4. Though the shot is fairly weak in color, using the Vectorscope as a guide, push the overall **Saturation** up a little. I used 20 taps on the **Up** arrow.
5. Now move to the **Color** tab. Switching back to the Waveform Monitor, display the **RGB Parade**. In the bottom of the three traces, we can see that the red channel is a little high in relation to the others. Move the **Shadows** puck to the left side and pull it down just a little to neutralize the blacks. I used 6^0, and went to -4% or 4 taps on the **Down** arrow.
6. Move the Highlights puck to the left side to yellow-orange around 32^0 and pull it down a little to about -13%.
7. Finally, warm up the mids with a little yellow, something like $+2\%$ at 52^0.

There is no right way to fix this as long as you set your exposure correctly, neutralize the blacks, and take the yellow out of the sky.

NOTE

Dutch

Dutch here means a shot that's tilted. The term goes back to the 1920s and is a corruption of *deutsch*, used because of German expressionist filmmakers' habit of shooting with a tilted camera to make shots more dynamic, intense, or disorienting.

Silhouette

Let's do a more complex image, with different problems, particularly in the light levels. Let's look at the *silhouette* clip in the project:

1. Put the playhead near the end of the clip when the woman holding the camera is pretty much in silhouette against the pale wall.
2. If necessary, press **Command-7** to open your scopes and from the **Settings** pop-up, select the Waveform Monitor displaying **Luma**.
3. Some like to adjust the overall puck and then fix the individual luma ranges, but I prefer to work with the individual pucks only. Start with **Shadows**, bringing them down a little to touch zero.

4. Next, pull down the **Highlights** a little to get the brightness under control and peaking at 100 IRE or 100%.

5. Then push up the **Midtones**. You can afford to push it up quite a lot to bring out the detail in the woman's face. I pushed it as high as 40% or 45%.

Here's a little problem this raises. Video cameras shoot what's called Y'CbCr. The b and r represent the blue and red channels, and the Y stands for the luminance channel, which is also the green channel. In most video editing applications, like previous versions of FCP, the application worked internally in Y'CbCr, just like your video. The current version of FCP works internally in RGB. What that means for us is that when we start to push the luminance values around, instead of being tied to a single channel, they're tied to all the channels, so moving the midtone luma as much as we are moves the color channels as well. To see that, switch the Display to the RGB Parade and start pulling the Midtones Exposure puck up and down. Watch what happens to the blue channel. This movement raises it significantly. Consequently, we have to make adjustments for this in the Color Board:

1. On our way to Color, first stop quickly in **Saturation** and push up the overall values a bit.

2. In the Color tab, take the overall puck and pull the blue down just a bit to take a little of the blue cast off the walls and the woman's arm.

You can tinker with this example for quite a while, but I think we should leave it there for the moment.

TIP

Applying a Filter to Multiple Clips

If you have a correction set that you want to apply to a number of clips with the same name in the Timeline, such as for an interview that recurs in the project, set the filter for the first clip and then copy it. In the Timeline, use **Command-F** just as in the Event Browser to search for all the clips with the same name. The Timeline Index opens for you to type in the name of the clip in the search box. This will filter all the clips in the Timeline with the same name. Select all the clips and use **Edit>Paste Effects** (**Option-Command-V**) to paste the same settings to all the clips with the same name.

Night

Let's look at the night shots. They are especially difficult because of the extreme low-light conditions. One of the problems, which really cannot be fixed, is that increasing the luminance levels will bring out a huge amount of grain. This can be improved to some degree by transcoding the media from the native AVCHD to ProRes like this:

1. Select the shots in the *night* Keyword Collection and use **File>Transcode Media**, or **Control-**click on the selected clips and choose **Transcode Media**.

2. In the dialog that appears, check **Create optimized media**.

As the material is converted to ProRes, the clips in the Timeline will be replaced with the ProRes files and used for the color correction. The grain will still be pretty bad, but with a little work, you might be able to save shots that are otherwise unusable. Let's start with the *ms musician* clip in the Timeline:

1. With the Waveform Monitor open, start in the **Exposure** tab by pulling down the blacks a drop or two so they're at zero or maybe even a touch below.
2. Leave the **Highlights** where they are or even push them a little beyond 100 IRE. These two will help to increase the contrast—blacker than black and whiter than white.
3. Use the **Midtones** puck to push up exposure to about +33% to bring something back to the picture.
4. Generally, night scenes are not heavily saturated, so we'll bypass Saturation for the **Color** tab. Night scenes usually have a slight bluish cast, a little colder than normal, so push the **Highlights** puck up about +20% right in the middle of pure blue.
5. Finally, take the **Mids** puck and put it right below the Highlights and pull it down just a bit, about –4%, to counter the highlight light and to take a little of the blue out of the musician's skin.

That's the image, perhaps as good as can be salvaged. What would help it, of course, would be a better camera, with higher resolution, that did not exhibit the tremendous amount of grain introduced in the low-light conditions.

TIP

Adjusting a Clip Before You Edit It

Sometimes you want to correct a clip before you put it into the Timeline. A shot such as an interview that's going to be reused multiple times might be better corrected before you edit it into the Timeline. To do this, select the clip in Event Browser, **Control**-click on it, and select **Open in Timeline**. In the Inspector, you now have access to the Color Board for Correction 1. Whenever you edit a piece of the shot from the browser into the Timeline, that color correction will be applied. Note that after you edit the shot into the Timeline, any subsequent changes made in the Event Browser to the color balance for that shot will have no effect on clips already edited into the project.

Broadcast Safe

Broadcast Safe is an effect that should be added to the *ms musician* shot. It's used to clip off any high luminance levels that appear on a clip.

1. Select the *ms musician* clip in the Timeline and open the Effects Browser with **Command-5**.
2. With All Video & Audio selected, type *broad* in the search box at the bottom. The Broadcast Safe filter will appear.

FIGURE 11.11

Broadcast Safe Controls.

3. Double-click this filter to apply the effect to the clip. If you have the Waveform Monitor open, you'll see the bright street lights that are exceeding 100 IRE being rounded off by the filter to 100 IRE.

The Broadcast Safe effect can be something of a magic bullet and works well at the default setting. In the effect controls (see Figure 11.11), you'll see you have separate options for reducing the luminance, the most common usage, or reducing the saturation. There is also a pop-up to select whether you're applying NTSC, the North American standards, or PAL, the European standards.

A common way to apply Broadcast Safe is to apply it globally to an entire project. You can either select all the clips and double-click the filter to apply it to everything, or you can make a Compound Clip. The first method is fine if you need only the default settings, but if you're working to PAL standards, you'll need to make a Compound Clip like this:

1. Select all the clips in the project and press **Option-G** to convert them to a Compound Clip.
2. With the Compound Clip in the Timeline selected, double-click the Broadcast Safe effect.

It will be applied globally to all the clips in the project. If you have the Waveform Monitor open and skim across the project, you'll see that the bright night shots that we haven't corrected will also have their luminance values chopped off so as not to exceed 100 IRE.

Generally, the process of converting the project to a Compound Clip and applying the Broadcast Safe effect is the last step before you export or share your finished work.

Match Color

The **Match Color** function in FCP is one of the great new features. Sometimes it just seems to have almost magical properties, although sometimes it doesn't work as well as you might like. In that case, you can use FCP's extensive color correction tools to fix the problem manually. There are four ways to apply Match Color:

• From the menus, select **Modify>Match Color**.
• From the Enhancements menu in the Toolbar, select **Match Color**.

- In the clip's Video tab in the Inspector under Color, click the **Choose** button next to Match Color.
- Use the keyboard shortcut **Option-Command-M**.

We've already fixed the medium shot of the musician as best we could, but there are two other night shots: a wider shot of the musician and a shot of a photographer with a decidedly greenish cast to it. Let's see what the Match Color function does for us:

1. If you have made the project into a Compound Clip to apply the Broadcast Safe effect, that's not a problem. Simply double-click the Compound Clip to open it into the Timeline.
2. Select the *musician night* clip and activate Match Color with one of the four ways just described.
3. A two-up display appears. The one on the right is the image you want to fix; on the left is whatever is underneath the skimmer.
4. Move the skimmer over the *ms musician* clip and click on it. The Match Color effect will match the shot on the right to the shot on the left (see Figure 11.12).
5. Click on **Apply** in the dialog to apply the selection.
6. Select the *night shot* with the green cast and activate Match Color.
7. Again skim over *ms musician*, click on it, and apply it to the green shot.

The effect is quite remarkable. The green is gone, and the image's luminance values match that of the first shot. Unlike the Broadcast Safe effect, the Match Color function is not so much of a magic bullet and sometimes doesn't work that well.

One thing you should be aware of is that the Match Color function does not actually use the Color Board to correct the color and luminance of the image. It simply makes mathematical calculations and applies them based on which frame you click in the sampling image, so nothing it has done is reflected in the Color Board. You may want to experiment by clicking in certain sections of the shot, such as the

FIGURE 11.12

Match Color Display.

beginning, middle, or end, for better matching values. Also, the Color Board controls are downstream on the Match Color function, which is probably a good thing because it allows you to make additional changes to what Match Color has already applied. For instance, I don't quite like what it did to the *musician night* clip. Let's fix it like this:

1. Open in the Color Board, and in Exposure, take down the black a little to bring it down to zero and even crush it a little.
2. In the Saturation tab, push the midtones up about 10% to bring more color back to the flesh tones.
3. Use the skimmer to compare the adjacent shots.
4. To return to the project from inside the Compound Clip, use the **back** button in the upper left of the Timeline or the shortcut **Command-left bracket**.

Looks

Looks in film and video are created by talented, creative colorists who bring magic to the movies. Some looks, such as Bleach Bypass, have become popular to the point of cliché. Some presets are provided as effects in FCP, as we have seen, but there are more available in the Color Board, and more importantly, you can make and save your own. Let's see how this process works:

1. You should be back inside the *Color* project with the single Compound Clip. As we're going to do more work on the project, let's break the Compound Clip apart. Select it and use **Clip>Break Apart Clip Items** or the keyboard shortcut **Shift-Command-G**.
2. Select the *silhouette with doors* shot and do a little basic correction to it in the Exposure tab of the Color Board.
   Pull down the **Highlights** a little, say −5%.
   Pull down the **Shadows** a little, about −1%.
   Push up the **Midtones** about +10%
   Increase the overall **Saturation** by +10%.
   In the Color tab shift the overall puck to yellow-orange in the image, about 38°, and increase by +5%.
3. After tweaking the image, go back to the **Video** tab of the Inspector and click the **+** button opposite Color to add a second color grade (see Figure 11.13).

Correction 2 is added beneath Correction 1. Generally, the basic color correction for an image—setting its luma and chroma levels, and matching it to other clips—should be done in one grade or correction, and any additional special looks or secondary color corrections should be done in additional grades that can be tweaked separately and switched on and off as needed. This also allows you to try different looks. Remember, you also have the Audition feature that allows you to make duplicate copies of clips to which you can apply separate looks and grade effects:

FIGURE 11.13

Adding Second Correction.

FIGURE 11.14

Color Looks Presets.

1. To apply the looks or to create your own, open the Color Board for Correction 2. Notice the pop-up in the upper-right corner that allows you to switch easily from Correction 1 to Correction 2.
2. To access existing presets, click on the gear pop-up menu in the lower right of the Color Board for a list you can choose from (see Figure 11.14).
3. Select **Summer Sun**. It's a warm, bright, high-contrast look.

Like most of them, the Summer Sun preset is extreme. It's designed for all material, so it uses extreme settings. All of them need to have their luminance levels fixed. The best way to fix these levels is to apply the Broadcast Safe effect, which clamps the levels without affecting the quality of the look across the central dynamic range of the shot. However, you cannot apply the Broadcast Safe effect to the clip if it has a second correction. You need to make the clip or the whole project with multiple corrections in a Compound Clip first, before applying Broadcast Safe. Applying the effect to the Compound Clip will clamp the whole image. This procedure works very well to rein in these extreme looks without ruining their effect.

Notice that in addition to the available presets, you can create your own by pushing the pucks around and then using the **Save Preset** selection in the menu. You get to name your preset, and it appears at the bottom of the preset menu. The presets are stored in your *Movies* folder in the *Final Cut Events* folder, where a folder called *Color Presets* is created to hold your favorite looks.

You can also get additional looks from companies like GenArts. Also, be sure to check out Denver Riddle's great Luster Grade presets at www.colorgradingcenteral.com.

SECONDARY CORRECTION

So far we've done only primary color correction, fixing the overall luma and chroma of an image, even adding looks that affect the whole image. Sometimes, however, you want to correct only part of an image—make the sky more blue; make the yellow more yellow; or highlight an area of an image, like an actor's face, to make it pop. This is done on most every shot in primetime television and in the movies.

Shape Mask

There are two ways to make a selection in FCP to restrict the correction we're making. We do this by making masks. A mask is the area we select that we want to control. A mask can be either a **Shape Mask** or a **Color Mask**. A Shape Mask is just that—a shape we create that defines the area of our correction—whereas the Color Mask is a specific color range that we select that we want to affect. You can use multiple shapes and color masks on every image, but let's try just a few and look at how these masks work.

Shape Masks are usually areas that are greatly feathered to gently blend the effect into the rest of the image. We'll start with the *cs statue* clip in the *secondary* Keyword Collection:

1. Append the clip into the *Color* project. Let's do a basic grade to adjust its levels in Correction 1.

   In the Exposure tab of the Color Board reduce the **Shadows** by −3% and the **Highlights** by −8%.

   Increase the overall **Saturation** by +20%.

FIGURE 11.15

Shape Mask Button.

FIGURE 11.16

Shape Mask Controls in the Viewer.

2. Return to the Video tab of the Inspector and add a second correction.
3. There are two buttons for creating masks. The one on the right makes a Shape Mask (see Figure 11.15), whereas the button with the eyedropper next to it creates a Color Mask. Click the Shape Mask button opposite Correction 2.

On-screen controls appear in the Viewer (see Figure 11.16). The inner circle is the area that will be affected by your controls. The outer circle is the fall-off from the control. You want to make the fall-off so smooth that the enhancement effect is not noticeable. The target in the center allows you to position the shape, while the handle allows you to change the angle. The dot to the left of the upper green dot lets you pull out the circle and change the shape to a rounded rectangle (see Figure 11.17). What

FIGURE 11.17

Rounded Rectangle Shape Mask.

we want to do is make a mask that allows us to brighten the statue while subduing the rest of the image so that the statue is the focus of the attention:

1. Draw out the top of the circle so it's a tall, narrow oval, with plenty of fall-off, and then position it around the statue.
2. Go to the Correction 2 Color Board and in the Exposure tab try these adjustments:
   Push up the **Midtones** about +15%.
   To increase the contrast, pull down the **Shadows** to −5%.
   Increase the **Highlights** by +6%.

This alters the area inside the Shape Mask. By increasing the contrast and brightness and, if there was color in the image, by increasing the saturation, you emphasize that part of the image.

To help the emphasis, you want to darken the outside, reducing the contrast and the saturation to de-emphasize that area. At the bottom of the Color Board, notice two buttons: one for Inside Mask and the other for Outside Mask (see Figure 11.18).

1. Click on **Outside Mask**. These are entirely separate controls, a completely separate Color Board from the one for Inside Mask.
2. To reduce the contrast, we'll push up the blacks and pull down the whites in Exposure.
   Push **Shadows** up +5%.
   Pull the **Highlights** down −5%.
   Pull down the **Midtones** a lot, to about −19%.

If the fall-off is sufficient, the brightening of the statue will be imperceptible, or at most, the audience will think what a fortuitous shaft of light fell on that statue at the moment.

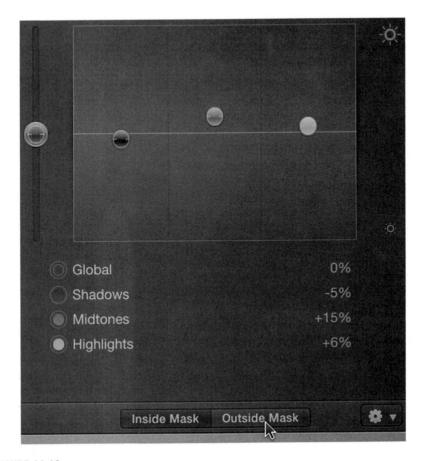

Global   0%
Shadows   -5%
Midtones   +15%
Highlights   +6%

Inside Mask | Outside Mask

FIGURE 11.18

Inside Mask and Outside Mask Buttons.

A commonly used technique with photographers is to use a graduated filter to reduce the brightness of the sky in the upper parts of an image, which can be overly harsh, drawing the viewer's eye to it rather than focusing on the subject. Let's do something similar to a portion of the *ws duomo* shot in the *secondary* keyword collection.

1. Select about the first four seconds of the *ws duomo* shot, before the camera pans left, and append that into the Timeline. The cathedral is too bright in relation to the rest of the image.
2. We'll dispense with the primary correction and go straight to the Shape Mask. Add the Shape Mask and drag out a highly feathered horizontal rectangle, as in Figure 11.19.
3. In the Exposure tab, pull down the **Shadows** just a little, down −4%. Pull the **Highlights** down just a little, about −2%. Pull down the **Midtones** to about −23%.

A separate Shape Mask could be used to reduce the brightness of the sky and the building on the right side of the frame.

FIGURE 11.19

Large Horizontal Rectangle Shape Mask.

NOTE

Animating Shapes

In these images we're working with statues and cathedrals, which don't move much, but sometimes you want to make a mask around a person, such as we did to the statue. People aren't so static, and Shape Masks need to be animated. This can be done by adding keyframes (see Figure 11.20) and manually repositioning and animating the motion as the mask needs to move around the screen.

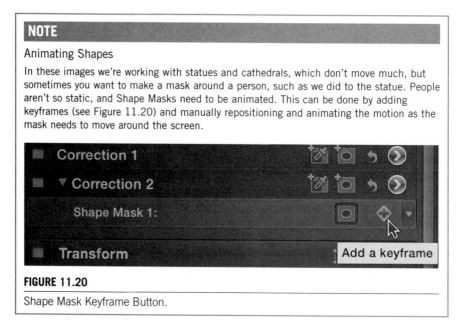

FIGURE 11.20

Shape Mask Keyframe Button.

Color Mask

Sometimes you don't want to adjust an area on the image; you want to adjust a color or range of colors, make a color pop, or take out the color from everything else in the frame. Fixing the color itself is the easiest part of the process. The hard part is

FIGURE 11.21

Using the Eyedropper to Make a Color Mask.

making the Color Mask and carefully selecting, adding to, subtracting from, and feathering the mask, like this:

1. For this exercise, we'll use the *duomo* clip from the *secondary* Keyword Collection. Append it into the storyline. What we want to do is to punch up the blue in the sky.
2. In the Video tab of the Inspector, click on the eyedropper button of Color Mask.
3. Go to the Viewer with the eyedropper and click and drag a small area of the sky (see Figure 11.21).
4. Hold the **Shift** key and drag another selection of the sky to add it to the selection. Repeat until you have most of the sky selected, but none of the building.
5. If you select too much, hold the **Option** key while dragging a selection to remove a color range from the selection.
6. In the Inspector, the slider between the eyedropper and the color swatch adjusts the mask feathering. Hold the **Option** key and drag the slider. While you're holding the key and dragging, you'll see a black-and-white representation of the Color Mask (see Figure 11.22). What's white is selected. What's black is not selected.
7. With the Color Mask selected in the Color tab of the Color Board, push the **Highlights** to blue at 239^0 and raising them +37%.

FIGURE 11.22

Mask Display in the Viewer.

8. In the Outside, set the overall color puck up in yellow a degree or two to warm it up and help the contrast with the blue of the sky.

That's it. The pale blue has become a vibrant blue that the chamber of commerce always wants.

Using these tools, you can see the great range of techniques that can be used and their power in shaping and controlling the way you use light and color in your productions.

SUMMARY

In this lesson we looked at the tools for color correction in FCP, starting with the automatic Color Balance function. For manual correction, we learned how to use the video scopes and the manual controls in the Color Board. We saw how to white balance clips and to fix a number of common color correction problems. We applied the Broadcast Safe effect to keep video from exceeding legal luminance levels. We also used the Match Color function to make two clips have the same color and tonal quality, and the preset looks to created special visual effects. Finally, we saw the power of FCP's multiple secondary correction tools, including the extraordinary control of Shape and Color Masks. Next, we'll look at the transform functions in FCP and use their animation capabilities.

Animating Images

12

LESSON OUTLINE

Final Cut Pro has considerable capabilities for animating images, allowing you to move them around the screen, resize, crop, rotate, and distort them at will. All of these properties can be animated, which allows you to enhance your productions and create exciting, interesting, and artistic sequences.

SETTING UP THE PROJECT

One more time, let's begin by preparing the material that you'll need. For this lesson we'll use *FCP7.sparseimage*. If you haven't downloaded the material, or the project or event has become corrupted, download it again from the www.fcpxbook.com website.

1. Once the file is downloaded, copy or move it to the top level of your dedicated media drive.
2. Double-click *FCP7.sparseimage* to mount the disk.
3. Launch the application.

The sparse disk has a project called *Motion* in its Project Library. We'll be working with this together with the event called *Motion* in the Event Library. Inside the event, you'll see the media we're going to work with. In this case it's entirely still images. We could use video clips, but for the purpose of this exercise, it's easier if the media is stills. Also we'll be working with still images to create motion on the photograph.

Before we begin, let's duplicate the *Motion* project:

1. In the Project Library select the project on the *FCP7* disk and use **Command-D** to duplicate it.
2. Leave **Duplicate Project Only** checked and uncheck **Include Render Files**.
3. Duplicate the project to your system drive or your media drive outside the *FCP7* disk. This will preserve the original for you to return to if necessary.
4. You can rename the duplicate project to something like *Motion copy* to distinguish it from the project on the disk image.
5. Open the duplicate project.

> **NOTE**
>
> Stills Frame Rate
>
> The project is 1280×720 pixels and is at 23.98 frames per second (more correctly 23.976fps, but we round it up when writing it). This frame rate is used by video when emulating film's 24fps rate. In the Events Browser, select the first image in Keyword Collection *720*. If you go to the Info tab of the Inspector, you'll see that the frame rate for the still image is listed as 60fps, the highest frame rate FCP utilizes. The still, however, will take its frame rate from the project in which it's placed. If you select the same image in the *Motion copy* project and check the Inspector, you'll see its frame rate has become 23.98.

TRANSFORM

FCP has three different functions for positioning and animating video or stills around the screen. The three functions are **Transform, Crop,** and **Distort**. There are several ways to access these functions:

- Using the buttons in the lower left of the Viewer (see Figure 12.1)
- **Control**-clicking on the clip in the Viewer itself and selecting the function from the shortcut menu (see Figure 12.2)

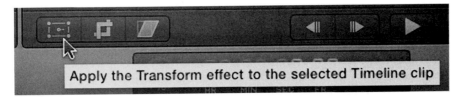

FIGURE 12.1

Transform, Crop, Distort Buttons.

FIGURE 12.2

Transform, Crop, Distort Shortcut Menu.

- Using one of the keyboard shortcuts: **Shift-T** for Transform, **Shift-C** for Crop, or **Shift-Command-D** for Distort
- Clicking on one of buttons for Transform, Crop, or Distort in the Video tab of the Inspector when the clip is selected in the Timeline

The easiest way to control these functions is directly in the Viewer with some handy on-screen tools. For precision, you also have access to numeric controls. You can find each of the functions in the Video tab of the Inspector and then double-click and open them to see the controls (see Figure 12.3). Using the on-screen controls together with the Inspector controls, you can manipulate your images with precision and ease. Furthermore, all of these properties can be animated; that is, you can add keyframes and make the transformations happen over time.

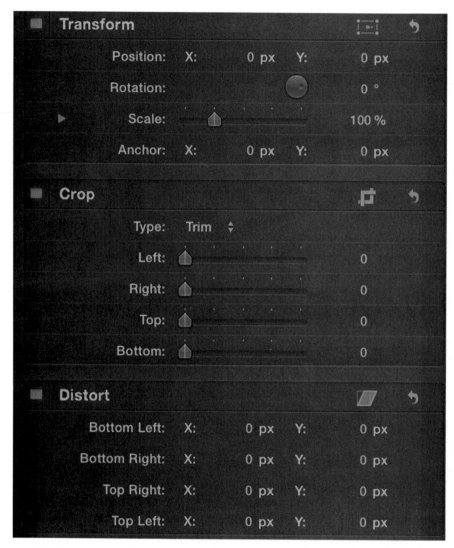

FIGURE 12.3

Transform, Crop, Distort Inspector Controls.

Scale

The first Transform function we'll look at is **Scale**. Here's how to use it:

1. Select the first image in the *Motion copy* project and activate the Transform function by pressing **Shift-T** or using one of the other methods. The image is outlined in white with round, blue dots on the edges (see Figure 12.4).
2. To scale the image, grab a corner and drag (see Figure 12.5). The image scales proportionately around the center.

FIGURE 12.4

Scale Outline.

FIGURE 12.5

Scaling an Image.

FIGURE 12.6

Changing the Aspect Ratio.

FIGURE 12.7

Scale Values in the Inspector.

3. If you drag one of the sides, the image's aspect ratio will change, squeezing the image and distorting it (see Figure 12.6). This scales the image separately in the X-axis and the Y-axis

4. If you hold the **Shift** key and drag a corner, you can scale and change the aspect ratio of the image at the same time, and these numbers will be reflected in the Scale values in the Inspector (see Figure 12.7).

5. If you hold the **Option** key and drag a corner, you can scale around any opposite corner, which remains pinned in its original position.

6. If you hold the **Option** key and drag a side handle, the image will stretch or squeeze from the opposite edge, which remains pinned.

You can scale an image larger than it is, but it's usually not a good idea to scale upward, not much above 110 to 120 percent. Higher than that, and the image will start to soften and eventually show pixelization.

Although the sliders and value boxes in the HUD give you precise control, the easiest way to scale or control the other Transform parameters is to use the Viewer.

Rotation

Rotation is controlled in the Viewer or with a dial and value box in the Inspector. There does not appear to be a limit to how many times you can rotate an image, which means if you animate the rotation value, the image can keep on turning. You rotate like this:

1. To rotate the image in the Viewer, grab the handle and pull (see Figure 12.8).
2. Hold the **Shift** key to force the rotation to snap to 45° increments.

The farther out you pull the handle, the easier it is to rotate with precision. With the handle close to the anchor point, which is in the center of the image, you can rotate very quickly.

Notice the dot on the right edge of the rotation circle. This indicates the last position of the image in its rotation. The first time you rotate a picture, it indicates what's

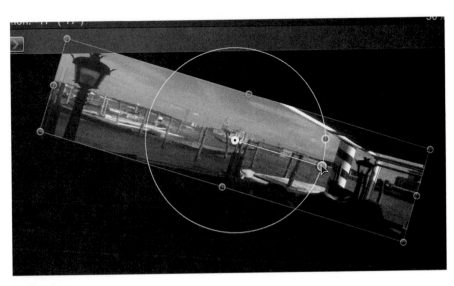

FIGURE 12.8

Rotating the Image.

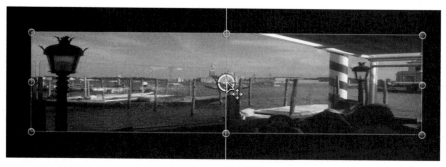

FIGURE 12.9

Positioning the Image.

horizontal, but the next time you rotate the image, it shows what its last angle was in the rotation circle.

Position

You can change the **Position** in the Viewer or with specific values for the X-Y coordinates in the Inspector. FCP counts the default center position (0, 0) as the center of the screen and counts outward from there, minus X to the left, plus X to the right, minus Y downward, and plus Y upward.

You can grab the image anywhere and drag it around the screen to position it, but if you drag it with the center target point, the image will snap to yellow center guides as you drag it (see Figure 12.9).

Anchor Point

The **Anchor Point** is the place on the image that everything moves around. The image scales to the Anchor Point; it rotates around the Anchor Point; and when you position the image, you are repositioning its Anchor Point on the screen. The default location of the Anchor Point is the center of the image, regardless of where the image is on the screen, but let's see what happens when it's moved. First, we'll reset the Transform function:

1. To reset the image back to its default state, click the hooked arrow opposite Transform in the Inspector.
2. Double click the X value of the Anchor Point, and using the scroll on your mouse, change it to –90. Alternatively, you can type in that value or mouse-drag within the X value boxes.
3. Change the Anchor Point Y value to 50. The Anchor Point is now in the upper-left corner of the image, which should be off the lower-right edge of the screen, as in Figure 12.10.
4. Grab the Anchor Button in the upper-left corner and drag the image to the upper-left corner of the Viewer using the alignment guides to help position the image (see Figure 12.11).

FIGURE 12.10

Changed Anchor Point.

FIGURE 12.11

Positioning with Alignment Guides.

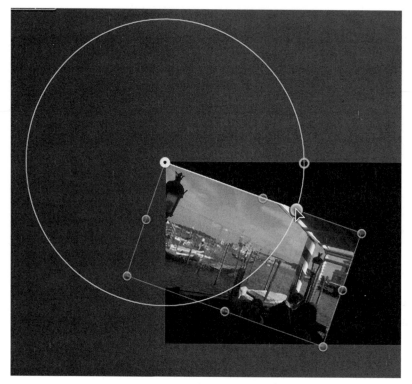

FIGURE 12.12

Rotating Around the Anchor Point.

5. Double-click the Scale value in the Inspector and change the value to 66%. The image scales around the upper-left corner.
6. The rotation handle is attached to the Anchor point, so it's also in the upper left. Pull it out and drag it downward. The image pivots around in the upper-left corner of the screen, as in Figure 12.12.

TIP

Precision Drag

To precision drag, hold the **Option** while dragging within any X, Y, or Rotation value box to increment change by tenths of a unit.

The Anchor Point becomes a powerful tool for animating images. After you finish working with the Transform function or any of the other two image manipulation functions, make a point of clicking the **Done** button in the upper right of the Viewer to close the function.

CROP

The **Crop** function allows you to trim the edges of a clip, cutting off the sides. You can use it to have multiple images on the screen or simply to remove an area that you don't want your audience to see. Be careful with this last function because it scales up the image, and that can make it look blurry or even pixelated, especially if it has to be recompressed for delivery on the web. There is a third crop function that creates a simple animated zoom on your image, a function that Apple calls the *Ken Burns effect*.

Trim

Let's leave the first image and work with the Crop tools on the second image, *0748*:

1. Select the image and click the **Crop** button, the middle of the three in the lower left of the Viewer. Three separate Crop functions can be selected using the buttons that appear at the upper left of the Viewer (see Figure 12.13).
2. Select the **Trim** function and notice the square corner tabs and the dotted line around the image.
3. To crop the image, drag a corner tab or one of the side tabs to pull in those edges and crop off part of the image (see Figure 12.14).
4. Once you've cropped an area, you can drag the cropped area around the screen to change the selection.
5. Drag with the **Shift** key to maintain the proportions. If you start holding the **Shift** key, the crop selection will remain 16:9. If you make the selection square and then hold the **Shift** key, the crop selection will remain square.
6. If you hold **Shift-Option** while dragging the corners of the image, it will be cropped proportionately on all sides.

You can also control the Crop function with precision use of the value boxes in the Crop section of the Inspector. Plus, you can switch the crop type from Trim to Crop to Ken Burns using the pop-up in the Inspector.

FIGURE 12.13

Crop Function Buttons.

FIGURE 12.14

Trimming the Image.

The area outside the cropped selection normally appears as black, but it's not. It's the emptiness of space. If you want the image to appear with a colored background, you should use one of the generators like this:

1. Select the image in the Timeline and use **Option-Command-Up** arrow to lift it out of the primary storyline so it's a connected clip.
2. Find the generator you want to put underneath it, perhaps Organic, and drag it onto the Gap Clip in the storyline. Then hold and select **Replace from start** from the shortcut menu.
3. Select the cropped image on top and click the **Done** button in the Viewer.

Crop

The **Crop** tool in the Crop function works a little differently. For one thing, no matter how you drag it, the image is always cropped proportionately. Why? Because when you finish cropping the image and click **Done**, the image will fill the screen.

Let's work on the third image in the project, *0755*:

1. Activate the Crop function with the button in the lower left and click the **Crop** button, the middle of the three in the top left of the Viewer (see Figure 12.15). Notice the image is outlined, but not with a dotted line; and now you only have drag tabs in the corners of the image.
2. Drag in the upper-right corner to tighten the image, cropping a little from the right and the top of the taxi driver's cap.
3. Grab the selected area and drag it up so there is a little more headroom.
4. Click **Done** and the image will fill the screen.

FIGURE 12.15

Cropping the Image.

Remember, this example scales the image up, so there will be some softening. Therefore, you should use this technique cautiously. It's a handy tool, though, if there's something on the edge of the frame that you need to cut out, like a microphone tangling into the shot.

If you wish, you can also drag beyond the edges of the image, which will create empty space around the image. This is an excellent way to give the image some breathing room around the image. And, of course, dragging it in this Crop function will create a proportionately equal amount of space around the whole image, though you could shift the crop selection off to one side if you wanted.

Ken Burns

While the **Ken Burns** function does crop your image, it is primarily an automatic animation tool. It's used in many Apple applications, iPhoto, and iMovie, and is used in the operating system to create small animations on screen-saver images. The effect produces a zoom into or out of the image. The duration and speed of the zoom are controlled by the length of the clip. The animation takes the whole length of the clip, so a short clip will have fast zoom, whereas a long clip will have a slow zoom. Let's apply the Ken Burns effect to the fourth image in the sequence:

1. Select the last image, *0778*, and click the Crop button or press **Shift-C** to activate the Crop function.
2. Click the third of the three buttons in the top right of the Viewer, the **Ken Burns** button. The image will appear in the Viewer with two boxes marked (see Figure 12.16,

FIGURE 12.16

Default Ken Burns Effect.

FIGURE 12.17

Ken Burns Buttons.

also in the color section), the green start box on the outer edge of the image, and the red end box about 15% into the image.

3. At the top of the Viewer, click the button with the two chasing arrows (see Figure 12.17). This swaps start for end. The effect will now zoom out.

4. Press the Play button next to the swap button to see the effect. It plays just the clip in Looped Playback mode. Click the button to stop playback.

5. To adjust the start position, you will need to move the end position because it completely covers the inner box. Push the red end box off the left.

6. Make the green start box smaller by dragging the lower-right corner inward and position the box so it's focused on the statue in the upper left.

7. Make the end box a little smaller and slide it back so the end position is completely over the image (see Figure 12.18). Notice the arrow that indicates the direction of the zoom motion.

FIGURE 12.18

Custom Ken Burns Effect.

8. If the motion is too slow, drag either end of the clip to make it shorter and speed up the effect.

9. Play the animation and click **Done** when you're finished.

Though you can zoom in further, especially if you're working with large-format still images, for video that matches your frame size, 15% is probably as far as you should go.

> **NOTE**
>
> Ken Burns Effect
>
> Although Ken Burns is a masterful storyteller and a great filmmaker, he did not invent the technique of moving across still images that he popularized in *The Civil War* television series, and in professional video production, it is not called the Ken Burns effect. The technique is called variously *motion control* or *pan and scan* or *image animation*. The term *Ken Burns effect* is used in iMovie and is just one of the terms that has been carried over from that application.

DISTORT

The **Distort** function will let you skew an image almost as if you're twisting it in 3D space. It behaves quite differently from the Distort tool in previous versions of the application. There the distortion was basically twisting the image in 2D space, while here it serves some of the functions of the Basic 3D filter in legacy Final Cut.

Let's reset the second image in the sequence, *0748*, and we'll look at the Distort function on that:

1. If you put a background underneath the image, collapse it back into the primary storyline with **Option-Command-Down** arrow.
2. Select the image and click the **Crop** reset button in the Inspector to reset it.
3. Click the last of the three transformation buttons, the **Distort** button. The image looks similar to the Transform function, except the drag tabs are square instead of round.
4. Grab tabs and start dragging a corner of the image to distort it (see Figure 12.19).
5. Drag the tabs on the edges to skew the image. The displacement will be reflected by corner point values in the Inspector (see Figure 12.20).

FIGURE 12.19

Distorting the Image.

FIGURE 12.20

Distorted Image and Corner Point Values.

6. You can drag the image to position it, although this is a Transform, but you do not need to switch to Transform to do it.
7. Again, when you're done, click the **Done** button.

MOTION ANIMATION
Straight Motion

Final Cut Pro allows you to create many different types of animation for both motion and effects. Many of the parameters in the Inspector can be animated. To make an animation, you have to understand the concept of keyframes, which may be new to iMovie users but should be familiar to Express and legacy Final Cut Pro users. When you change the properties of an image, you define how it looks. When you apply a keyframe, you define how it looks at a particular moment in time, at a specific frame of your video. If you then go to another point in time, some other frame of video—say, five seconds farther into your video—and change the values for the image, you can define how it looks at that point in time. The computer will then figure out—it will *tween*—what each of the intervening frames of video would look like over that five seconds. If the keyframes are far apart in time, the change will be gradual. If the keyframes are closer together, then the change will be more rapid. We've already done some animations in which we animated audio levels. Now let's do a simple motion animation.

It's easy to set a motion keyframe in FCP. There is a keyframe button right in the Viewer. Let's start by creating a simple animation on the last image, *0078*:

1. Start by resetting it by clicking the **Distort** reset button in the Inspector.
2. Next, let's set the duration of the image. Select it, press **Control-D**, and type *500* and press **Enter** to make it five seconds long.
3. In the Viewer, go to the **Zoom** pop-up menu and set it to display the screen at 50% or a value that will give you some grayboard around the edge of the screen. You can go down to 25% in the pop-up, but using **Command-minus** in the Viewer will let you go down to 15% and **Command-plus** up to 3000%, too!
4. Scale the image down to 50%. You can switch on Transform and drag the corner of the image or type *50%* in the Transform Scale value in the Inspector.
5. With the Transform function activated, drag the image so it's in the top-left corner of the screen, as in Figure 12.21.
6. Move the playhead so it's one second after the beginning of the clip in the Timeline and click the keyframe button in the upper left of the Viewer. This sets a global keyframe at the point for all the Transform values, Position, Rotation, Scale, and Anchor Point.
7. Go to the one second point before the end of the clip in the Timeline.
8. Drag the image across the screen so it's in the lower-right corner of the screen, as in Figure 12.22.

FIGURE 12.21

Start of the Animation.

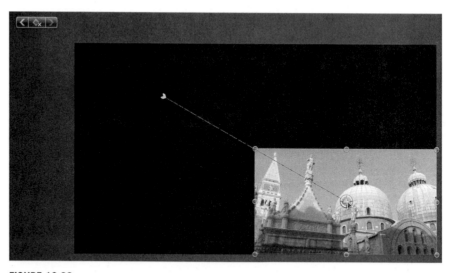

FIGURE 12.22

End of the Animation.

Notice two things: (1) the red dotted line appears on the screen indicating the motion path; and (2) the keyframe button has gone orange, indicating that another keyframe has been added automatically. The other thing you may notice as you play the animation is that it has *easing*. This is default behavior for all animations in FCPX. Easing means that the animation accelerates and decelerates—it eases into the motion and eases out of the motion; it does not simply stop and start.

That's it; you've created your first motion animation with keyframes. The possibilities are endless. To see what you can do with these tools, watch the opening titles for some television shows and see how motion graphics animations can be used to enhance your work.

It is very important to note the difference in keyframe behaviors between animating motion using Transform, Crop, or Distort in the Viewer and animating in the Inspector. In the Viewer, global keyframes are set for the properties for the motion functions, and any time you move the playhead and change the values of the motion function, new keyframes are automatically added. This was the way earlier versions of FCP handled keyframing. However, in the Inspector, keyframes are handled the way they are in the Motion application; you must manually set a keyframe for each property you want to change every time you want to change a value. There is no automatic keyframing of values in the Inspector. Even if you are animating the Transform, Crop, or Distort functions in the Inspector, they will still not be automatically keyframed. The automatic Viewer keyframes work only when the animation is controlled in the Viewer.

To delete a keyframe, you can go to keyframe with the arrow buttons next to the keyframe and then click the keyframe to remove it (see Figure 12.23).

If you want to move the keyframes to speed up or slow down the animation, it's easiest to do in the Video Animation controls (**Control-V**) in the Timeline. Select the parameter whose keyframes you want to alter (see Figure 12.24), and then drag the keyframes farther apart or closer together.

FIGURE 12.23

Deleting a Keyframe.

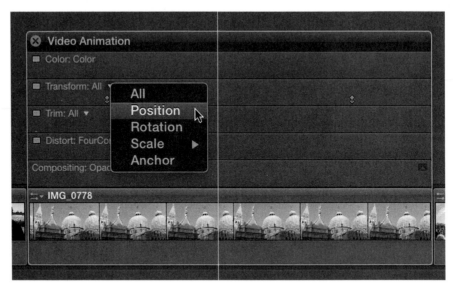

FIGURE 12.24

Video Animation Keyframes in the Timeline.

Curved Motion

You can create a curved path in two ways: (1) pull out the path from the linear motion or (2) create a curved path by using Bezier handles. In the first method, you move the playhead to the midpoint of the curved motion and drag the image out so that the path is curved (see Figure 12.25). This creates a new keyframe and gives the motion path a natural curve. The second method doesn't create an intermediate keyframe. If you go to keyframe points, you can **Control**-click on the keyframe and select **Smooth**. Bezier handles that control the motion path will appear; you can pull them out to change the shape of the path (see Figure 12.26). Unfortunately, when Smooth is used to create the motion path, easing is switched off.

TIP

Starting Off-Screen

If your animation starts off the screen, generally, you use the Smooth selection to create the Bezier handle. This adds the handle but no easing. The assumption is that the object arriving from off-screen is already in motion at full speed and does not accelerate as it enters the frame.

FIGURE 12.25

Curved Motion Path.

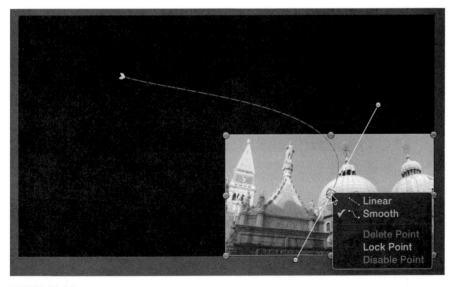

FIGURE 12.26.

Using Bezier Handles.

SPLIT SCREEN

A split screen is desirable for many reasons: for showing parallel action, such as two sides of a phone conversation, or showing a wide shot and a close-up in the same screen. It's easy to do if the video was specifically shot for a split screen. A phone conversation, for instance, should be shot so that one person in the conversation is on the left side of the screen and the other person is on the right side.

Let's do a simple split screen using the images we have in the project, but first let's reset them all:

1. Select all the images, and in the Inspector, click the reset buttons for **Transform**, **Crop**, and **Distort**.
2. Let's stack two images on top of each other. Take the third image, *0755*, and drag it on top of the second, *0748*, so they are connected. With snapping turned on, the image should snap right into place. The magnetic timeline will close up the gap so your project should look like Figure 12.27.
3. With the Connected Clip selected, activate **Crop** and select **Trim**.
4. Grab the right side tab and drag it to the middle. It should snap to the center line.
5. Select the image on the primary storyline and click on **Transform** to activate that.
6. Slide the image to the right and frame it as you like.
7. A separator bar is really useful for split screens. Go to the **Generators** and select one of the **Solids**. The Custom Solid gives you the most freedom.
8. Drag it on top of the other two images so it snaps to them as a third Connected Clip.
9. Trim off the end by dragging it so all three images are the same length.
10. Select the **Solid** and activate **Crop** and the **Trim** function.
11. Crop the left and right sides and change the Solid color to whatever you like in the Generator tab of the Inspector. Your project and Viewer should look like Figure 12.28.

Picture in Picture

By now, you've probably figured out how to make a Picture in Picture (PIP). You just scale down the image to the desired size and position it wherever you want on the screen. One note of caution about PIPs: Many video formats, such as material converted from analog media, leave a few lines of black on the edges of the frame.

FIGURE 12.27

Project Layout.

FIGURE 12.28

Split Screen in the Viewer and Timeline.

These are normally hidden in the overscan area of your television set and are never seen. However, as soon as you start scaling down images and moving them about the screen, the black line becomes apparent. The easiest solution is to take the Crop tool and slightly crop the image before you do your PIP so you don't get the black lines, which give the video an amateur look.

You might also want to add a border to the PIP to set it off, either with the Simple Border effect, or if you want something fancy, you might try the Frame effect. Be careful, however, if you need to crop the image. The cropping will cut off the border.

One way around this is to put a generator background between the PIP and the underlying video and crop the generator to fit around the PIP as a border. To keep them together, you can select the two items and make them a Compound Clip.

Multilayer Animation

Let's build a multilayer stack in FCP to make what's called a *Quad Split*, and then we'll animate it. A Quad Split is four images on the screen, one in each corner. It's a little tricky to build a stack of clips that are already in the project because of FCP's magnetic timeline, but let me show you how. Start by deciding what you want to be the bottom layer in the stack. Let's say you want the last image, *0778*, as the bottom layer. You have two images stacked on top of each other next to it. In that case, you cannot simply pick them up and move them on top of the other image. Instead, you do the following:

1. Reset the image. Select all four images and click the **Transform**, **Crop**, and **Distort** reset buttons to make sure they're back to normal.
2. Pick up the first image in the project, *0375*, and drag it on top of the last image so it's a Connected Clip.
3. Drag out the end of the image on the primary storyline so the two images are the same length.
4. To move the other two images, you have to lift them out of the primary storyline. Select the image on the storyline, *0748*, and use **Option-Command-Up** arrow to lift it out.
5. Drag the two Connected Clips and drop them on top of the last two clips so your project looks like Figure 12.29.
6. Select the Gap Clip at the beginning of the sequence and delete it.

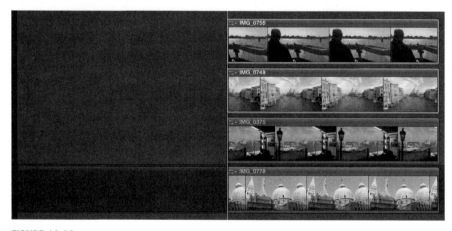

FIGURE 12.29

Stacked Project.

Now you have the stack built, it's time to create the Quad Split:

1. Select all four images in the project, and in the Transform section of the Inspector, dial in a **Scale** value of **50%**. All the images will be resized to the same dimension in the center of the Viewer.
2. Deselect everything so you can move the images individually.
3. Activate the **Transform** function (**Shift-T**) and drag the top image to the upper-left corner of the Viewer.
4. Move the second image to the upper left, the third to the lower right, and the fourth to the lower left so the Viewer looks like Figure 12.30.
5. To animate the images, move the playhead to the two second point in the Timeline and select all the images with **Command-A**.
6. In the Inspector, click the **Position**, **Rotation**, and **Scale** keyframe buttons. This is the end position for your animation. You cannot click the keyframe button in the Viewer to keyframe multiple items.
7. Move the playhead to the beginning of the project and click the same keyframe buttons to add another keyframe for the start of your animation.
8. Now scale and spin and scatter the images to the four corners. With all the images selected, your Viewer might look something like Figure 12.31.

That's it. Play your animation to see your Quad Split reassemble itself on the screen over the first two seconds.

FIGURE 12.30

Quad Split.

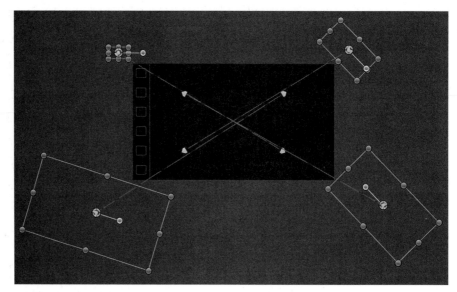

FIGURE 12.31

Images Off-Screen.

TIP

Resetting Clips

Another way to reset clips is to select one that is normal, like a clip in the Event Browser, and copy it. Then select the clips you want to reset and use **Option-Command-V** for Paste Effects to paste the default normal state to the other clips.

MOTION CONTROL

Motion control is another term for the Ken Burns effect, also called *pan and scan.* It's a technique that's very useful on large-size still images. It has always raised some issues working in Final Cut Pro and still does in the current version. While the Ken Burns effect using the Crop function works well, it is limited to animation that you want to last the duration of the clip. Often, however, you'll want to hold on an image and then push in or pull out and then hold again at the end. Often you'll want to make animations on still images or large graphics such as maps that are quite a bit larger than your frame size. Let's see what happens in FCP when we do a large-scale animation.

In the *Motion* event, find the *0773* image and append it into the project. By default, the image will be made to fit the project. If you look in the Info tab of the Inspector, you'll see that the image is 2592×1936 pixels in size, much larger than the size of the project we're working in, which is 1280×720 pixels. This is no more

---

TIP

Viewer Navigation

If you zoom into the Viewer for a detailed look of your media, a little box with a red selection area will appear (see Figure 12.32). This is a navigation aid. You can grab inside the box and drag the red selection area around the screen to navigate to different sections of the Viewer.

FIGURE 12.32

Viewer Navigation.

---

than a medium resolution image. These days, it is not uncommon for images to be 4288×2848 pixels or greater. Whenever you bring a still image into an FCPX sequence, the application will always try to fit the image to the sequence size as best it can. If the image is tall and narrow—a vertical shot—the application will scale it so that it fits vertically in the frame. In this case the image is a different aspect, so it shows pillarboxing—black bars on the left and right side of the image. To see the image at normal size:

1. Select the image in the Timeline, and in the **Video** tab, find **Spatial Conform** and set the pop-up to **None**. The image will now fill the frame.
2. Activate the **Transform** function to see the size of the image in relation to the screen (see Figure 12.33). To create the motion control effect, we'll need to animate the Transform function.

FIGURE 12.33

Spatial Conform Set to None.

FIGURE 12.34

Position for Start of Animation.

3. Move the playhead about two seconds in from the start of the clip and click the keyframe button in the Viewer.
4. Move the image in the Viewer as far down in the frame as it will go so that the top edge is resting on the edge of the screen, as in Figure 12.34. This is our start position.
5. Move forward in the Timeline about five seconds.
6. Drag the corner of the image to scale it down till the sides align with the sides of the Viewer. Then slide the image up so the top of the image remains aligned with the screen.

7. Play the animation. You'll see that the motion does a slight S-curve, dipping into the frame and then coming back again.
8. Go back to the start keyframe and adjust the start position to get a little wiggle room so the image does not move completely off the screen.

This movement during the zoom occurs because the motion animation of scaling is different from that of position. Position is a linear animation, while scaling is logarithmic. When an image is at 100%, it fills the screen, but when it's at 50%, it's only a quarter of the size of the screen. FCP corrects this issue to some extent, which is why you see only a slight snaking during the zoom. This problem was much more pronounced in earlier versions, but it's still not completely cured, and you need to be careful when using this technique.

Another way to do this animation is to use the Crop function. It makes a clean motion, but it looks a little mechanical because it has no easing; plus, it's difficult to calculate how far into the image you can zoom because you have no value box that displays how much the image is being scaled when it's cropped. Let me show you how to construct this:

1. Start by appending another copy of *0773* into the Timeline. Leave its **Spatial Conform** as **Fit**. If you change it, you'll get unexpected results.
2. Move the playhead to about three quarters of the duration of the image. This will be your end position.
3. Activate **Crop** and select the **Crop** button, the middle of the three buttons at the top of the Viewer.
4. Move the selection area to the top of the screen so the top edge lines up with the top of the Viewer, as in Figure 12.35.
5. Click the keyframe button in the Viewer to set the end point for the animation.
6. Move the playhead to about one second after the start of the image.
7. Drag the corner of the image to reduce the size of the selection. How far you can drag it is pretty much guesswork. Make it small enough so the winged lion is centered in the frame and clearly visible as the focus of interest.
8. When you have it cropped and the selection positioned as you like, click **Done**.

The animation will start zoomed in on the winged lion and pull back to as far as it can go within the image without showing any black around the edges.

Which option you use probably depends on the situation. Sometimes one might work better than another. For smaller movements, simply scaling and positioning will work fine, but for larger movements, you may want to resort to the Crop function to eliminate the weaving.

TIP

Checking the Scale Size

If you are going to animate the Crop function, it's a good idea to set the Spatial Conform of the image to None to see what kind of magnification limits you have when you're at 100%. This will give you an indication of how far you can scale the image. After you've assessed the image, set Spatial Conform back to Fit before you begin your Crop animation.

FIGURE 12.35

Crop Selection for End of Animation.

SUMMARY

In this lesson we looked at Final Cut Pro's animation capabilities. We used the Transform function to scale, rotate, and position the image. We also looked at Anchor Point behavior and how it affects an image's motion. We cropped the image using the Trim function to resize the image, the Crop function to reframe the clip, and the Ken Burns feature to create simple animations. We distorted the image and then animated it on the screen, in both straight and curved motion paths. We also looked at making basic motion effects like split screens, Picture in Picture, and multilayer animations. Finally, we looked at the problems of motion control across large format images. In the next lesson we'll look at more compositing techniques.

Compositing

13

LESSON OUTLINE

Compositing is the ability to combine multiple layers of video on a single screen and have them interact with one another. This capability adds great depth to FCP. In the previous lesson on animation, we built stacks of images in layers. Compositing allows you to create a montage of images and graphics that can explain some esoteric point or enhance a mundane portion of a production.

Good compositing work can raise the perceived quality of a production. Compositing is used for a great deal of video production work on television—commercials, of course—but also on news programs and for interstitials, the short videos that appear between sections of a program. Be warned, though, that compositing and graphics animation are not quick and easy to do. Most compositing is animated, and animation requires patience, skill, and hard work.

SETTING UP THE PROJECT

For this lesson we'll use *FCP8.sparseimage.* If you haven't downloaded the material, or the project or event has become messed up, download it again from the www .fcpxbook.com website.

1. Once the file is downloaded, copy or move it to the top level of your dedicated media drive.
2. Double-click *FCP8.sparseimage* to mount the disk.
3. Launch the application.

In this lesson we'll be working with the media called *Composite* in the Event Library.

BLEND MODES

One of the best ways to combine elements such as text or generator items with images is to use *blend modes.* If you're familiar with Photoshop, you probably already know that a blend mode is a way that the values of one image can be combined with the values of another image. Final Cut has 26 blend modes, including stencils and silhouettes. Any video element in the Timeline can have a blend mode applied to it. For graphical elements like generator items or titles, Compositing is the first item in the Video tab of the Inspector. For video clips, it's down at the bottom of the list together with the Opacity slider, which is a type of compositing mode. From the Blend Mode pop-up, you get to select the compositing type you want (see Figure 13.1). The default, of course, is Normal, which is the way the clip appears without any blend mode applied. The blend modes are listed in groups. The best way to see the blend modes is to try them out. Before you start to do heavy text and compositing work, I think it's a good idea to switch off background rendering. You don't really need your system constantly rendering as you make changes. Most often the material will play in real time. There may be occasional dropped frames, but they're not a great inconvenience and playback does not stop. To switch off playback rendering, go to **Final Cut Pro>Preferences,** and in the Playback tab, uncheck the box for **Background render**.

To help you see what the blend modes do, I've created a project with the same clip on it and the same underlying generator repeated again and again. Each copy of the clip has a different blend mode applied to it. The lower third on each clip identifies the blend mode applied to the video clip in the stack. I left out a few that have specific uses and wouldn't show anything if composited on video.

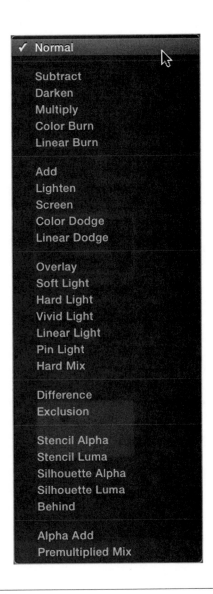

FIGURE 13.1

Compositing Blend Modes.

Compositing Preview

Let's begin by looking at the blend modes available in FCP. Open the project titled *CompositeModes*.

The first group of blend modes darkens the image, and the second group acts to brighten the image. Some, such as Screen and Multiply, are used to create transparency in an image. The next group affects primarily the contrast of the image, and these modes are not unlike lighting effects. Difference and Exclusion have some of the most pronounced effects on the image. The transparency group of stencil and silhouette are used to affect the transparency of underlying images. The Behind mode simply moves the image below the underlying layers. As you look through the project, you'll see that some of the differences, especially within each group, can be pretty subtle. It's a wonderful tool for controlling and combining images, especially text over video. To look at some of the blend modes, especially their specific uses, we'll start by making a new project on the system or media drive and naming it *Composite*.

IMPORTANT NOTE

Additional Storylines

Though the compositing blend modes appear for the clips in an additional storyline, they do not blend with the layers beneath it. If a clip or group of clips get converted to an additional storyline, the blend modes applied to those clips will disappear. This makes applying transitions between Connected Clips that have blend modes applied virtually impossible.

Screen and Multiply

One of the most useful blend modes is **Screen**, which will remove black from an image. It screens out portions of the image based on luminance values. Pure black will be transparent, and pure white will be fully opaque. Any other shade will be partially transparent. This is great for creating semitransparent shapes that move around the screen, and it is useful for making animated backgrounds. Many 3D and other animation elements are created against black backgrounds that can be screened out. Let's try the blend mode, and you'll see how it works, compositing the Lens Flare generator on top of another scene:

1. Let's put in the background layer first. From the *Composite* Events Browser, find the *Kabuki3* shot.
2. Press **Option-2** to do a video-only edit. We don't need the sound for these compositing exercises.
3. Append the shot into the *Composite* project.
4. Move the playhead about halfway through the shot, and in the browser, find the *LensFlare* shot.
5. Connect the shot into the project and play it. All you see in the middle of the shot is the lens flare moving across the black screen. The underlying shot is hidden.

6. Select the *LensFlare* shot in the Timeline, and from the bottom of the Video tab of the Inspector, click **Show** to open the controls. Then change the Compositing Blend Mode pop-up to **Screen**. Voila! We see the flare moving across the image.

7. Next, move the playhead back to the beginning of the project and connect the *Multiply* graphic to the project.

8. Select *Multiply* in the Timeline and change its Compositing Blend Mode to **Multiply**.

Instantly, the white background will disappear, and the black text will appear over the video and the lens flare.

These two effects can be seen in the assembled project called *BlendingFun* on the *FCP8* disk image. A number of the other effects we do are also available in this project.

NOTE

Lens Flare Size

The *LensFlare* movie is a little undersized, and you can see the edges at the brightest part. Not a problem! Just use the Transform function to scale up the image a little.

Animated Text Composite

One of the best uses of blend modes is with text, especially animated text that's moving on the screen. The blend mode will allow the text to interact with the layer underneath, giving a texture to the text. We're still working with the *Composite* project primary storyline.

1. To start, let's trim off the excess of the *Multiply* graphic so it's the same length as the *Kabuki3* shot.

2. Next, append *Village3* into the project to be our background for the text.

3. Because FCP has so many prebuilt text animations, we'll not use our own animation, just use one of the title templates. Put the playhead a short way into the *Village3* shot and connect the **Drifting** title on top of it.

4. For this type of thing, we want to use a large, chunky font and make it a bright color. I like Arial Black for this. In the Title tab of the Inspector, set both lines to this font.

5. Set both Line Sizes to **250**, big and bold, and pick a fairly bright red for both.

6. For the text, type the word *DAMINE* in the first line. It's the name of the village, and for the second line, type *JAPAN*. To keep the bold effect, type in all caps.

7. Finally, in the text's Video tab of the Inspector, right at the top, change the compositing blend mode. I used **Color Dodge** in Figure 13.2 (also in the color section).

These kinds of compositing techniques are used constantly in motion graphics. Even if you don't use a blend mode, simply adjusting the opacity of your text so the face has a little transparency will create interesting effects.

This effect is also in the *BlendingFun* project.

FIGURE 13.2

Text with Color Dodge.

NOTE

Unrendered Playback

If you have background rendering switched off while you do compositing work, you might notice a little wobbling or shimmer as you play the titles back. The reason is that the computer is playing back the unrendered title and trying to produce it in real time. To render the title, select it and press **Control-R** for render selection. Playback will be a lot smoother now.

TIP

NTSC Warning

If you are going to output to NTSC analog to be seen on a television set, be careful in using blend modes, particularly Add. It will brighten the image, often beyond the luminance and chrominance values allowable for broadcast transmission. If the image is too bright or oversaturated, especially in red, it may bloom objectionably and smear easily when converted to formats such as MPEG-2 for DVD.

Adding a Bug

A *bug* is an insect, a mistake in software coding, and also that little icon usually in the lower-right corner of your television that tells you what station you're watching or who the program producer is. There are lots of different ways to make bugs, but I'll show you one using Photoshop and blend modes. In your *Composite* Event Browser is a Photoshop file called *Logo*. The file is actually made up of two layers, one of which was switched off in Photoshop. Layered PSD files appear as flattened images in FCP, though the transparency is maintained. To start, do the following:

1. Put the playhead at the beginning of the *Composite* project and connect the *Logo* file into Timeline.
2. Note the duration of the project at the bottom of the Timeline, select the *Logo* clip in the project, and open the **Duration** box in the Dashboard by double-clicking it.
3. Make the duration the same as the duration of the project. Mine is 21:07. Your project should look like Figure 13.3.
4. With the *Logo* selected, go to the Video tab of the Inspector, and in Spatial Conform, set the pop-up to **None**. It's still much too big for a bug, but let's work on compositing it first.
5. **Control**-click on either the *Logo* in the browser or in the project and select **Reveal in Finder (Shift-Command-R)**.
6. Open the file in Photoshop, and with the layer selected in the Layers panel, click the little *fx* at the bottom left and choose **Drop Shadow** (see Figure 13.4).
7. Leave the **Drop Shadow Angle** at **145** but set the **Distance** to **15**. Leave the other controls the same.
8. Select **Inner Shadow** and check the box to add the shadow. Leave the **Distance** at **5**, but set the **Choke** and **Size** to **10**.
9. Also add a **Stroke**. Set the **Size** to **6**. I made the color blue.
10. Click **OK** and save the file. You've built the effect.

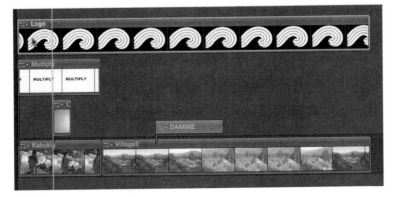

FIGURE 13.3

Project in Timeline.

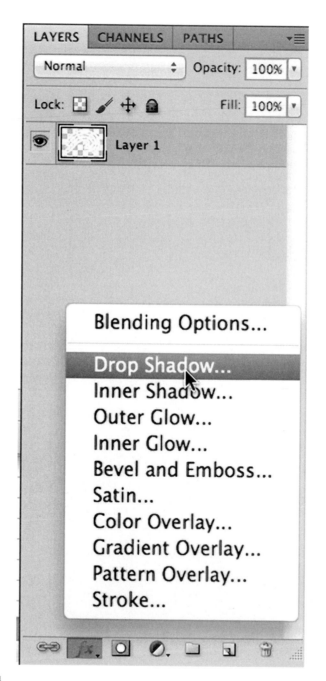

FIGURE 13.4

Adding Drop Shadow in Photoshop.

When you switch back to FCP with **Command-tab**, the Photoshop files will be updated, both in the Event Browser and in the Timeline.

The next step is to change the logo's composite type. Several blend modes will work for this, but I like to use Multiply because this will make the white of the logo almost transparent. For a slightly brighter look, try Soft Light, and for an even less transparent look, use Overlay.

At the current size, the logo is probably a bit intrusive. Use Transform to scale it down a bit and reposition it in the corner of your choice. The bottom right is the traditional location on American television, but I like the bottom left for this. It will now be your unobtrusive watermark on the screen (see Figure 13.5). Some people also like to use effects such as displacements or bump maps in Photoshop, which add texture, but for something this small, I don't really think these effects are necessary. The simple transparency effect of a blend mode is enough.

I left this effect off the *BlendingFun* project so that you have the opportunity to build the logo for yourself in Photoshop. If you don't have Photoshop or Photoshop Elements, you can use the *Logo2* file in the Event Browser to substitute for the bug.

FIGURE 13.5

Bug on the Screen.

> **TIP**
>
> Distinctive Logo
>
> There are any number of different ways to give your logo a distinctive edge by using the power of layer styles in Photoshop. Rather than using the Inner Shadow and Stroke method, you could also use Bevel and Emboss. Change the Technique pop-up to Chisel Hard and push the Size slider until the two sides of the bevel meet. This will give you maximum effect. Remember that the logo is going to be very reduced in size. Also, in the bottom part of the Bevel and Emboss panel, try different types of Glass Contour from the little arrow pop-up menu.

STENCILS AND SILHOUETTES

Travel mattes, or traveling mattes, are a unique type of composite that uses the layer's transparency or luminance to define the underlying layer's transparency. The old film term *travel matte* is no longer used in FCP, where we now have *Stencil* and *Silhouette* blend modes. These modes allow the application to create a great many wonderful motion graphics animations. We'll look at a few that I hope will spur your imagination and get you started on creating interesting and exciting projects. Technically, these are blend modes as well, even though they function in a special way. There are two Stencil modes—one for **Luminance** and one for **Alpha**—just as there are two Silhouette modes—one of each type. In these modes, the layer to which the blend mode is applied will give its shape and transparency from either the Luminance value or the Alpha (the transparency) value of the layer to the layers beneath it. A Stencil mode on a layer will reveal only what's inside the selected area, whereas a Silhouette mode on a layer will show what's outside the selected area, basically punching a hole through the layer or layers underneath it. In the project, any animation or change in the layer with the blend mode will be reflected in the layers below it. This makes Stencils and Silhouettes extraordinarily powerful tools.

These concepts are harder to explain in words than they are to demonstrate, so let's make a new project that we'll use to see how these compositing tools work:

1. In the *FCP8* disk image, find the project called *TransparencyModes* and duplicate it to your system or media drive.
2. Name the new project *TransparencyModes copy*.
3. Open the project, select the *Stencil* text clip on top, and in the Video tab of the Inspector, set the Compositing Blend Mode to **Stencil Alpha**.

    All you see is the bamboo and the grass (see Figure 13.6). The text itself is gone. Its only purpose is to provide the transparency shape information for the layers below it.
4. Select the *Stencil* layer and the *Natural* layer beneath it, and convert them to a Compound Clip with **Option-G**. Now you see only half of the word because the layer with the grass is cropped.
5. Undo that and select *Stencil, Natural,* and *Wood* and convert them to a Compound Clip.

FIGURE 13.6

Stencil Alpha Layer.

FIGURE 13.7

Stencil Alpha in Compound Clip.

Now you see the word divided into wood and grass, but the Stencil Alpha mode affects only those two layers, and for the first time, you can see the *Metal* layer beneath (see Figure 13.7).

Let's look at what happens with the Silhouette Alpha mode:

1. In the *TransparencyModes copy* project, select the *Silhouette* clip and in the Video tab of the Inspector, change its Compositing Blend Mode to **Silhouette Alpha**. You will see the bamboo and the grass with the word *SILHOUETTE* punching a hole through the layers (see Figure 13.8).
2. Select the top three layers—the text, the wood, and the grass—and convert them to a Compound Clip. You'll see the text has cut a hole, but only through the first two layers, revealing the metal plate underneath (see Figure 13.9).

Here, we are using just text, but you can use the transparency channel of any image to create interesting effects. I think you can see the potential of this powerful compositing tool here.

FIGURE 13.8

Silhouette Alpha Layer.

FIGURE 13.9

Silhouette Alpha in Compound Clip.

The Stencil and Silhouette Luma modes work similarly. They use the luminance information of an image to either cut out the area surrounding the stencil or cut through the area of the silhouette. Let's try the Stencil Luma on the first stack:

1. Double-click the Compound Clip in the first stack to open it into the Timeline.
2. Select the text layer and in the Inspector change the blend mode to **Stencil Luma**.

The luminance value of the text is being used, and because it is not fully white, it will show some transparency. Only what is pure white will be opaque, and pure black (or transparent) will be transparent in the underlying layers.

Switch back to the *TransparencyModes copy* project, and you will see the stencil effect with the word ghosted on top of the metal background. If the red of *STENCIL* were pink or paler, the text would be more distinct on the metal (see Figure 13.10). Make the Color adjustment in the **Inspector>Text>Face** section and see what happens. Try also the Silhouette Luma mode on the Silhouette Compound Clip.

Organic Melding

A common technique that always works well is to use luma maps to create effects that allow images to meld into each other organically. Usually, this is done by creating a grayscale image with motion in it that is used as the luminance map. Let's do this in the *Composite* project that we were working on earlier:

1. To build this effect, start by making sure you're set to do a video-only edit (**Option**-2).

FIGURE 13.10

Stencil Luma in Compound Clip.

2. Next, append the *Kabuki1* clip to the end of the project. It's only five seconds long, so, if necessary, zoom into the Timeline (**Command-plus**) to see it more easily.
3. Move the playhead back to the beginning of *Kabuki1* and connect *Archer1* to the storyline.
4. Drag the end of the Connected Clip to shorten it to the length of the underlying clip.
5. To add the luma map, we'll use the Underwater generator. Move the playhead to the beginning of the stack and connect *Underwater*.

> **NOTE**
>
> Effect and Generator
>
> Don't confuse this with the identically named Underwater effect under Effects. If you can't connect it, it's an effect. Look for the same name under Generators, which can act like clips.

Before we composite, there's a small problem we have to take care of first. The Underwater generator is quite long and the speed of the watery motion is determined by the length of the clip, so we don't want to shorten it. That would make motion very quick and not at all what we want. There is a way around this. Basically, we "nest" the generator in a Compound Clip all its own, so its motion happens there, and then we shorten the Compound Clip like this:

6. Select the *Underwater* clip and press **Option-G** to make it a Compound Clip.
7. Shorten the Compound Clip to make it the same length as the other two clips in the stack.

Next, we need to create a luma map out of this. To do this, we'll adjust the Compound Clip luminance values, remove the color, and increase the contrast so that the motion has more effect on the images and gives a better definition to the effect:

8. Select the Compound Clip and press **Command-6** to go to the Color Board.
9. In the Saturation tab, pull the master puck all the way down to **0** (zero).

10. In the Exposure tab, set the **Shadows** down to **−33%**, the **Midtones** down to **−68%**, and to increase the contrast, push the **Highlights** up to **+38%**.
11. Go to the Compound Clip's Video tab and set the Compositing Blend Mode to **Silhouette Luma**.
12. Finally, to complete the effect, to blend the Compound Clip with the base layer, select the *Underwater* Compound Clip and *Archer1*. Then use **Option-G** to combine them into a new Compound Clip, a Compound Clip within a Compound Clip.

What immediately happens is that you see the underlying *Kabuki1* clip and the two blend together based on the luminance and the motion of the Underwater generator (see Figure 13.11).

This is in the *BlendingFun* project on the disk image.

Highlight Matte

Next, we're going to create a **Highlight Matte**. This allows us to create a highlight area, like a shimmer that moves across an image or, as here, across a layer of text. Let's set up a simple animation:

1. Edit only five seconds of the video of one of the clips into the Timeline to use as a background layer. I used five seconds of *Ceremony 1*.
2. Next, we'll add text on top of it. Use the Custom Text tool, which is the basic text tool in FCP, and connect it to the beginning of the *Ceremony1* clip.
3. Change the text to the word *JAPAN* in any font you like, in a fairly large size, and in a nice, bright color. I used Optima for this, ExtraBlack, with a size of 220 in a fairly

FIGURE 13.11

Silhouette Luma Effect.

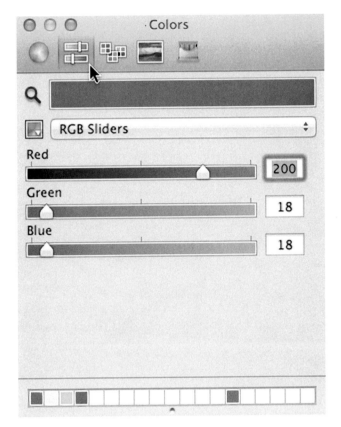

FIGURE 13.12

RGB Sliders.

bright red, R 200, G 18, B 18. To set the text color accurately, click on the color swatch and switch the Apple color picker display to RGB sliders, as in Figure 13.12.

4. Next, duplicate the text block and make a connected copy in the space above by **Option**-dragging the text upward.

5. Open the top text clip, and make it a very pale version of the text color beneath it. I made it a pale pink.

6. Move the playhead back to the beginning of the stack, select **Shapes** from the Generator Browser, and connect it on top of the other clips. Make it the same length as the other clips.

Now let's work on the highlight:

7. In the Generator tab of the Inspector, set the **Shape** pop-up to rectangle.

8. Uncheck the **Outline** and set the **Drop Shadow Opacity** to **0** (zero).

9. In the Video tab of the Inspector, open the **Transform** section and set the **Rotation** value to **90°**.

FIGURE 13.13

Viewer with Highlight Bar.

10. In the Effects Browser in the Blur group, find the **Gaussian** blur and apply it to the Shape.

11. Set the **Gaussian Amount** to **100**.

Your Viewer should look something like Figure 13.13.

Animating the Highlight

The next step will be to animate the highlight. We'll do this in the Viewer with the Transform function:

1. Select the *Shapes* clip and put the playhead at the start of the clip.

2. Drag the shape so it's off to the left, past the edge of the text, and click the keyframe button in the Viewer to set a Transform keyframe.

3. Move to the last frame of the *Shapes* clip and drag it across the screen so it's clear of the right side of the text. Click **Done** to complete the animation.

Over the five seconds of the clip, the Highlight bar will sweep slowly across the screen. Of course, we still don't see the background layer.

4. Next, set the Compositing Blend Mode of the *Shapes* clip to **Stencil Alpha** or **Luma**. In this case, they will produce the same result.

5. Finally, select the *Shapes* clip and the upper text file and make them into a Compound Clip.

If you're at the start or end of the clip, you'll see the red text that's directly above the *Ceremony1* clip as well as the background, of course. As you skim over the stack, you'll see that the text highlight area will softly wipe onto the screen and then wipe off again as the Highlight layer slides underneath it. The pale text file's transparency is directly controlled by the transparency value of the layer above it. The matte layer, the *Shapes* clip in this case, is invisible. This highlight animation and the others below are in the *BlendingFun* project for you to check out.

Glints

We've seen how we can put a highlight across an image. Next, we're going to do something a little more complex—create a traveling highlight—but one that goes only along the edges of a piece of text, a highlight that glints the edges.

1. We'll begin again with a base layer from one of the clips in the Browser. I used another five-second piece of *Ceremony1* for this, appending it into the primary storyline.
2. Next, we'll add the Custom text above it. Copy the *JAPAN* text clip that's above the first *Ceremony1* clip and paste it on top of the second one.

Next, we'll create a moving highlight area:

3. From the Generators Browser, we'll select the **Shape** generator again and connect it at the beginning of the second ceremony clip and on top of the text, making it the same length as the others.
4. In the Generator tab of the Inspector, switch off the **Outline** and set the **Drop Shadow Opacity** to **0** (zero) again.
5. We need the Circle to be much smaller, so set the **Roundness** value to about 130.
6. We also need it blurred, so use the **Gaussian** blur effect at the default setting, which is fine for this.
7. We also need to move it higher in the frame using the Transform function. Set it up so the soft-edged circle affects only the top part of the text, as in Figure 13.14. Don't forget to click **Done**.

Animating the Circle

To animate the circle, we want it to move from left to right across the screen and then back again, keeping it at its current height. At the start, we want the circle off to one side of the screen.

1. Move the playhead to the beginning of the clip, and using the Transform function, move the circle straight to the left so it's clear of the text.
2. Set a Transform keyframe in the Viewer at the Start of the clip.

FIGURE 13.14

Circular Highlight Above Text.

3. Switch back to the Timeline and press **Control-P** to type in *+215*. Don't overlook the plus sign. Deselect the clip and then press **Enter** to move the playhead forward two and a half seconds, halfway through the clip.

4. Move the circle straight across to the right side so it's clear of the text on the right.

5. Go to the last frame of the five-second clip and move the circle back across the screen to where it came from.

Over the five seconds of the clip, the circle will sweep across the text and then back again. So far, so good. If you skim the Timeline or play through it, you should see the circle swing from left to right and back again.

6. To make the glow appear only on the text, start by **Option-**dragging the text clip up, allowing the Magnetic Timeline to push the circle out of the way onto a higher level. Your stack should look like Figure 13.15.

7. Select the top text layer, and in the Inspector, use the Text controls to change its Face color to white, pale yellow, or whatever glow color you want to use.

8. To the circle *Shape* clip, apply the **Stencil Alpha** Compositing Blend Mode.

9. Finally, select the *Shape* clip and the *JAPAN* text directly beneath it. Then combine into a Compound Clip with **Option-G**.

Immediately, the circle will disappear, and the glow will be composited on top of the lower text layer. The highlight areas move across the top edge of text. The result is similar to what we did before, but let's expand on that.

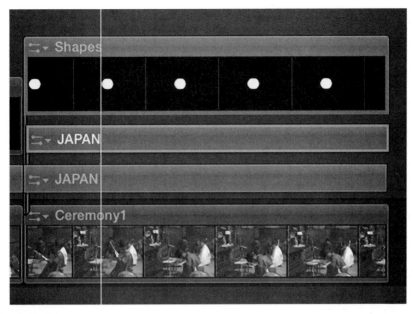

FIGURE 13.15

Graphics Stack.

Polishing the Glow

What we really want is for the glow not to race across the whole text, but to run just along the top edge of the text. It's not hard to do. We just need to add a few more layers:

1. **Option-**drag two more copies of the text layer situated under the Compound Clip to the very top of the stack, completely hiding the glow.
2. Select the topmost text layer, and in the Video tab of the Inspector, set the **X Transform Position** value to **0.5** and **Y** value to **–0.5**, half a pixel to the left and half a pixel down.
3. To the topmost text layer, apply the **Stencil Alpha** Compositing Blend Mode and combine the two top text layers into a Compound Clip.

The glow now runs on the little glint on the edge. You could do this just by putting an offset layer on top. It works, but it makes the font a little chunky looking, with the extra work to limit the glint to the area within the text. Using the approach here, you're left with just a glint that travels along the edges of the text (see Figure 13.16).

One More Touch

At the moment, the glint brushes across the upper-left edge of the letters as it moves back and forth across the screen. If you want to be really crafty and add a little something special, you can shift the glint side as it swings back and forth. For the

FIGURE 13.16

Text with Glint.

first pass of the blurred circle that creates the text glint, leave the settings as they are. Then do this:

1. When the glint reaches the far-right side of the screen at 2:15, we'll shift the position of top layer of the text that's inside the upper Compound Clip. Break apart the top Compound Clip with **Shift-Command-G**.
2. With the playhead halfway through the stack, 2:15 from the beginning, select the top text layer, the one that's offset. Then go to the Video tab of the Inspector.
3. Click the Transition **Position** keyframe button to keyframe that one parameter.
4. Go forward one frame in the Timeline and add another **Position** keyframe.
5. Set the **X** value to **–0.5**, moving it one pixel left, while leaving the **Y** value unchanged.
6. Select the two text layers and collapse them again into a Compound Clip with **Option-G**.

Now when the glint passes the first time from left to right, it will appear on the upper-left edge of the text, and when it goes back from right to left, it will be on the upper-right edge of the letters (see Figure 13.17).

Video in Text

I hope you're getting the hang of this by now and are beginning to understand the huge range of capabilities that these tools make possible. Next, we're going to make an even more complex animation with traveling mattes, the ever-popular video-inside-text effect, the kind of technique that might look familiar from the opening of the old television program, *Dallas*.

To look at what we're going to do, open the *MasterTitle* project on the *FCP8* disk image. That's what we're going to build. It might look like only two layers, but in fact it's much deeper and more complex than that.

FIGURE 13.17

Text Glint on the Right Edge.

Getting Started

To start, we'll edit a clip into a new project to establish the frame resolution and timebase. Then we'll build on that:

1. On either your system drive or your media drive, make a new project and call it *MasterBuild* or whatever you like.
2. Make sure video-only edits are switched on with **Option-2**. We're going to be editing a lot of clips into the project, and we don't need sound from them.
3. Append the *Kabuki1* clip into the project.
4. Use **Option-Command-Up** arrow to lift the clip from the primary storyline. This will leave the base storyline free for the background video, and the Gap Clip you create will act as a placeholder that can be replaced.

Making Text

Next, we need to make the text. The text is going to be filled with video and will fill the screen, so the letters need to be very large, as large as we can make them without crowding the edge of the frame. You don't need to worry about the title safe area because the text will be animated across the screen.

1. Put the playhead back at the beginning of the Timeline and double-click to connect the **Custom** title into the project. It's conveniently the same five seconds as our video clip.
2. Double-click the text to select it and type in the word *JAPAN* in caps.
3. In the Text controls in the Inspector, set the font to **HeadlineA**, set the **Size** to **615**, and set the **Baseline** to **–180**. The text is the right size horizontally, but we really need to make it bigger vertically.
4. We'll use the Transform **Distort** function (**Shift-Command-D**) to do this.
5. Zoom out in the Viewer using **Command-minus** to give yourself more room to drag the Distort controls.

FIGURE 13.18

Distorted Text.

6. Drag the top and bottom Distort handles up and down to increase the size of the text. It's very important that the distortion is only vertical and there is no horizontal distortion. When you have it right, click **Done**. Your text should look something like Figure 13.18.

Separating the Letters

You now have the base layer for the text. It's very important that you do not move the text, change its position or size on the screen. We're going to make multiple copies of the text that all have to line up exactly:

1. Holding the **Option** key, drag a copy of the text directly above it to make a new Connected Clip.
2. Repeat the process until you have a total of six text layers, one for each letter of the word and one extra (see Figure 13.19).

FIGURE 13.19

Text Stacked in the Timeline.

3. To separate the text, we're going to crop each word except for one letter. Start by selecting all the text layers except the bottom one and pressing the **V** key to make them invisible.
4. Adjust the Viewer so you have a little grayboard around the screen and activate the **Crop** function (**Shift-C**), making sure it's set to the **Trim** function.
5. Drag the right-edge crop handle to crop all the letters except the first *J*.
6. With the first layer selected, press **V** to hide it and then select the second text layer and press **V** to reveal it.
7. Crop from the left and right so that only the first letter *A* is visible.
8. Hide the second text layer and reveal the third.
9. Crop from both sides so that only the *P* is visible.
10. Repeat for the next layer, hiding the other layers, revealing the fourth layer and cropping to show the second *A*.
11. Finally, with only the fifth layer visible, crop from the left side only so that the final *N* is visible. Click **Done** when you're finished with the Crop function.
12. Leave the top layer hidden for now. We'll come back to that at the end.

Adding Video

We're now ready to put in the video. We're still working with video only here—no audio. To simplify this process and make it easier to see what's happening, we'll start by working with one text layer and then add in each one sequentially till each of the letters is filled with a different piece of video. We already have the first piece of video for the first letter in the Timeline.

1. Use the **V** key to switch off all the text layers except the first.
2. Select the *Kabuki1* clip and switch on the Transform function (**Shift-T**).

3. Scale the image, distort it, and position it as you like so that it covers the area of the letter *J* (see Figure 13.20).
4. Select the text layer *J*, and in the Video tab of the Inspector, set the Compositing Blend Mode to **Stencil Alpha**.
5. Select the two layers, the *Kabuki* video clip, and the text layer with the letter *J* and combine them into a Compound Clip with **Option-G**.
6. With the Compound Clip selected, go to the Info tab in the Inspector and change its name to *J*. Otherwise, all your Compound Clips will be named *Compound Clip*.

That's the basic process. We just have to edit in more clips and repeat the transforming and compositing process four more times:

FIGURE 13.20

Video Ready to Be Composited with First Letter.

7. Move the playhead back to the beginning of the project and connect the *Kabuki3* clip into the Timeline. (You need to trim one frame off the end of it to line it up with the other clips.)

8. Drag the clip down so it's below the second text layer. This procedure is tricky because of the nature of the Magnetic Timeline and takes some practice. Drag it to a position similar to Figure 13.21, and once it's on the right layer, slide it to the left to the beginning of the sequence.

9. Turn on the *A* layer. Then scale and distort and position the video layer to fit under the letter. Try to keep the girl's head in the top part of the *A*.

10. Apply **Stencil Alpha** to the text layer and then combine the two layers into a Compound Clip, renaming it *A*.

11. For the letter *P*, we'll connect the *Archer1* clip to the beginning of the Timeline and trim the end to match the length of the other clips.

12. Transform the video clip to be scaled and positioned directly beneath *P*.

13. Apply **Stencil Alpha** to the text layer, make it a Compound Clip with the video, and name the Compound Clip *P*.

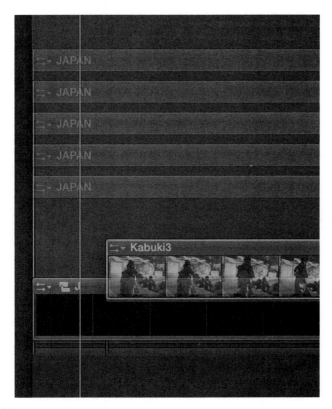

FIGURE 13.21

Positioning the Clip.

14. Now we're on a roll. Make the next text layer visible and connect the *Kabuki2* clip to the head of the project, moving it beneath the second *A* layer.

15. Set the second *A* layer to **Stencil Alpha**, combine it with the video, and name the Compound Clip *A*.

16. Turn on the letter *N* text layer, and connect and position the last five seconds of *Archer2* underneath it.

17. Apply **Stencil Alpha** to the *N* layer and combine it with the *Archer2* layer, naming the Compound Clip *N*.

Congratulations! You've got through the hardest, most tedious part of building this composition. Your Timeline should look like Figure 13.22. You've composited multiple tracks of video with multiple tracks of text, blending the two together to create a composition that shows an array of images simultaneously in the screen within the constraints of the text.

We're almost there—just one more refinement before we animate the image.

Edging

The last thing that works well on this type of composition is to add an outline edge. It helps to separate the text from the underlying image. We'll add an edging that borders the letters and separates the text and the video from what will be the background layer. That's what that last layer, which has been hidden for so long, is for.

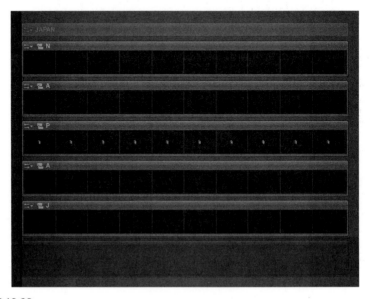

FIGURE 13.22

Project Timeline After Compositing Video.

1. Make the top text layer visible. It should completely cover all the underlying layers.
2. Select the layer, and in the Text tab of the Inspector, open the **Face** controls and set the **Opacity** to **0** (zero).
3. Turn on the **Outline** function and set the **Width** to about **6**. The default red works well for this. The Viewer should look like Figure 13.23.

Animating the Text

Before we animate the composition, we should put in the background video. This is what the Gap Clip we created when we lifted out the first video clip is for. We'll use *Village1* for this.

1. Find *Village1* in the Event Browser, drag it onto the Gap Clip to do a Replace edit, or press **Option-R** to do a Replace from Start edit.
2. Select all the Connected Clips and Compound Clips, everything above the primary storyline. Then press **Option-G** to convert it into a single Compound Clip.
3. Set the Viewer to a comfortable size that allows you to have some grayboard around it and then move the playhead to the beginning of the project with the **Home** key.
4. Select the Compound Clip, switch on the Transform function (**Shift-T**), and drag the Compound Clip straight off the right side of the frame, as in Figure 13.24. In the Inspector, check that the Position Y value is still at 0 (zero).

FIGURE 13.23

Composition with Edging.

FIGURE 13.24

Compound Clip Ready to Be Keyframed.

5. Click the Transform keyframe button in the Viewer to set a keyframe for the Transform functions.
6. Move to the last frame with the **End** key and then use the **Left** arrow to step back one frame.
7. Drag the Compound Clip straight across the screen till it disappears off the left side, again making sure that the Inspector still has the Position Y value set to 0 (zero).

Over the course of the five seconds, the text and the video inside it will travel from right to left over the top of your background video. There is a huge amount you can do with these tools. Play with them and have fun!

SUMMARY

We're now at the end of this lesson on compositing. We looked at Final Cut Pro's compositing tools, including a prebuilt preview to show off the various effects. We worked with the Screen and Multiply blend modes because of their unique properties. We also built a text composite using these tools and added a bug to our project. We then worked with some of the most powerful compositing blend modes, the Stencil Alpha and Luma and the Silhouette Alpha and Luma functions, which allowed us to use one image to cut through multiple layers of video. It also has the ability to let clips meld organically with each other. We created highlight mattes, animated them, and made them into glints, which we also animated. Finally, we built a complex multilayer project of video within multiple letters of a title, all animated against a video background. With that, we're almost ready to export our material from Final Cut Pro out into the world, which is the subject of our final lesson.

Outputting from Final Cut Pro

14

LESSON OUTLINE

Remember I said the hard, technical part of nonlinear editing was at the beginning, setting up and setting preferences, importing, and logging; the fun part was the editing in the middle; and the easy part was the outputting at the end? Well, we're finally up to the easy part: the output. Basically, when outputting, you're delivering your media in some digital form, whether it's heavily compressed for low-bandwidth web delivery or a high-resolution, uncompressed HD or higher format; you're delivering a digital master.

SETTING UP THE PROJECT

As always, begin by loading the material you need on the media hard drive of your computer. For this lesson we'll use *FCP9.sparseimage*. If you haven't downloaded the material, or the project or event has become messed up, download it again from the www.fcpxbook.com website.

1. Once the file is downloaded, copy or move it to the top level of your dedicated media drive.
2. Double-click *FCP9.sparseimage* to mount the disk.
3. Launch the application.

The media files that accompany this book are heavily compressed H.264 files. Playing back this media requires a fast computer. If you have difficulty playing back the media, you should transcode it to proxy media.

The sparse disk has a single project in its Project Library. We'll be working with this as the project we'll be outputting.

SHARE

With the exception of the XML export function, which is in the **File** menu, all of FCP's other outputting options are under the **Share** menu. I know the **Share** name rankles many users because it has a California hippy, touchie-feelie quality to it, but that's where the product came from, after all, and it is an attempt to smooth the transition for iMovie users to a professional application by keeping a common nomenclature. Whatever you call it, this is the place where you export your finished project. A project can be shared either from an open Timeline or directly from the Project Library. Any selected project can be exported, except for the **Save Current Frame** function, which can be done only from inside the Timeline window. At this writing, there is no batch export function; only individual items can be exported. That said, you do not have to wait for each export function to be completed before initiating another export. The export function is usually handed off to another application—the **Share Monitor**—to handle it, and FCP can, in fact, be closed.

TIP

Sharing from the Event Browser

While you cannot share directly from the Event Browser by selecting an item, you can **Control**-click on an item and select **Open in Timeline**. Once the item is in the Timeline panel, it can be shared just as if it were a project.

Media Browser and Apple Devices

The **Share** menu is broken into groups (see Figure 14.1). The first three groups are really the sharing functions in FCP because they are designed for direct export to specific designations. The export functions are below that and allow exporting to any place on your hard drives you want.

Exporting to the **Media Browser** creates a file or multiple files that are stored with your project file, in the project folder, inside a folder called *Shared Items*, and that can be accessed by other Apple applications using their Media Browser functions, applications like iDVD, iWeb, GarageBand, and so on.

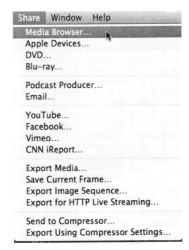

FIGURE 14.1

Share Menu.

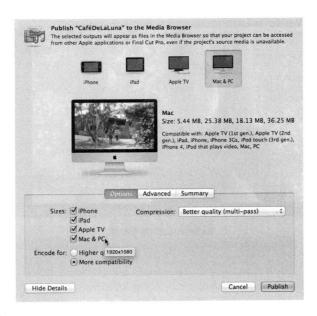

FIGURE 14.2

Media Browser Dialog.

Opening the Media Browser function brings up the dialog in Figure 14.2, which is the expanded dialog. In the basic dialog with **Show Details** closed, you can select to which Apple-compatible device you want to export. Though these are specifically designed for Apple device specifications, the exported file can be used for whatever you want, and the Mac & PC file is a standard MPEG-4 file that can be put on the web, though it is usually a pretty large file. When you select one of the four options, you get a little display of the device that you can skim through to see your project, just in case you selected the wrong one.

The **Show Details** button allows you to share to multiple devices at one time. You also have the option to check **More Compatibility**, which gives you specifications for older versions of Apple TV and older iPhones. These files will be more heavily compressed and smaller for these poorer-quality output devices. You also have the option to change the **Compression** pop-up to **Better quality (multi-pass)**, which is useful if your material has a lot of action, fast cuts, or complex effects and transitions.

The **Advanced** tab (see Figure 14.3) gives you access to Compressor 4's multi-machine processing capabilities. This requires having multiple copies of Compressor installed on multiple machines, and it also requires you to use the **Send to Compressor** function to share your project. Compressor 4 is a separate product currently sold for US $49.99 on the App Store.

The **Summary** tab (see Figure 14.4) simply gives you details of the file formats you're sharing. When you're ready to go, click the **Publish** button.

The **Apple Devices** option in the **Share** menu is exactly the same as the Media Browser option, except that instead of putting the exported files in the *Shared Files* folder associated with your project, a pop-up lets you pick whether to move the output file to your iTunes library or to a specific iTunes playlist. This is a good place to export your media if you're planning on syncing it to one of the Apple devices.

NOTE

Compressor 4

Though Compressor 4 is worth a slim book of its own, I've put together a basic lesson for working with Compressor on my web site that you should find useful. It doesn't require downloading any additional media, and will give you a overview of the application and get you started working with it. You can find it at http://www.fcpxbook.com/Compressor.

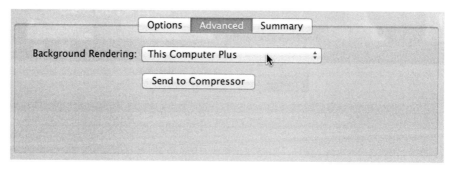

FIGURE 14.3

Advanced Tab.

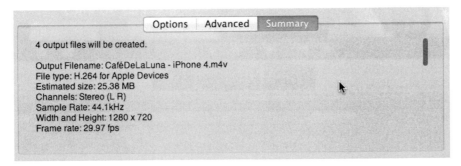

FIGURE 14.4

Summary Tab.

FIGURE 14.5

Exported Project Icons.

After you've shared or exported your project, a marker appears in the Project Library next to the project, indicating it's been shared. If you make a change to the project after you export it, the icon appears with a warning indicator to show the project has been changed. The two icons can be seen in Figure 14.5. In the

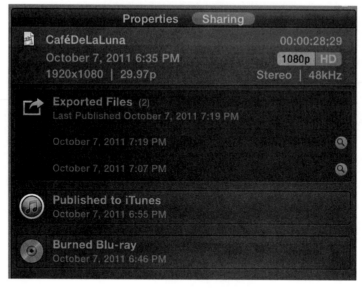

FIGURE 14.6

Sharing Tab in Project Properties.

Sharing tab of the Project Properties (see Figure 14.6), you can see what's been output and even click the magnifying glass on some of the files to point to where the files are.

Share Monitor

As soon as you send off your request to process your media, a separate component called the **Share Monitor** opens (see Figure 14.7). This feature allows you to monitor the progress of the media that's being encoded.

In the toolbar at the top of the Share Monitor, you have buttons to **Expand** and **Collapse** the processes displayed in the Share Monitor. On the right, you can sort by user and whether you want to display **All** the processes, **Active** processes, those being currently encoded, or **Completed** processes.

In the sidebar on the left, you can select any computer that is being used for processing. The reason this is here is that the Share Monitor is also employed by Compressor, which allows distributed processing across multiple computers. The main panel on the right displays the processes as selected in the toolbar.

On the right side, opposite each file, are three buttons: one that will pause an encode; one that will stop the process; and the little *i* button that will give you additional information about the process, elapsed time, remaining time, and in the case of finished processes, how long it took to encode the material (see Figure 14.8).

FIGURE 14.7

Share Monitor.

FIGURE 14.8

Information Detail.

Blu-ray and DVD

While it's Apple's view that silver discs, CDs, DVDs, and Blu-ray discs, while, if not entirely dead, are fast dying out as a delivery mechanism, nonetheless many people still want to make them, play them, buy them, and sell them. Apple has provided a very minimal mechanism for doing this within the application. If you want fuller features for standard-definition DVD creation, you would need to use iDVD or DVD Studio Pro, if you can find the latter somewhere. For more complete Blu-ray features, you would need Roxio's Toast or Adobe's Encore.

In the **Blu-ray** Share option (see Figure 14.9), the first pop-up lets you select whether to create an AVCHD red laser-burned disc, which burns onto a standard DVD, does not hold as much (usually less than an hour), and plays on most Blu-ray players; or a Blu-ray-encoded disk image onto your hard drive that can be burned to disc later. If you have a Blu-ray burner attached to your computer, that option will appear in the pop-up.

The second pop-up lets you select whether you're creating a single-layer or dual-layer disc. The pop-up below allows you to choose from one of five basic menus (see Figure 14.10). You can also name the disc and set whether the disc displays a menu first, or is what's called a *first play disc*, in which the program begins playing as soon as the disc is loaded into the player. There's also a checkbox that allows you to make the movie play in Looped mode.

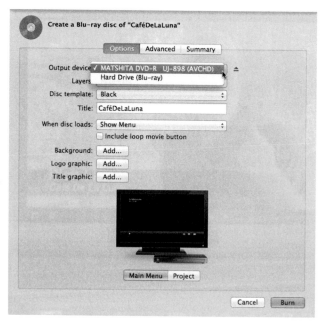

FIGURE 14.9

Blu-ray Share Option.

FIGURE 14.10

Blu-ray Menu Options.

If you have prebuilt graphical elements created in an application like Photoshop, you can choose to load those as a background image, a logo image, or as title text.

The Advanced and Summary tabs are the same as in other modules; they give you access to Compressor rendering and provide summary information for your output file.

When you click **Burn,** the process is sent to the Share Monitor until it's completed, at which time another application will open and ask you to insert a disc.

I think it's generally good practice to create the disk image first, saving it to your hard drive, and burn the necessary discs from that using the Disk Utility or Roxio Toast.

The standard-definition DVD-burning module is very similar to Blu-ray, except you have only two menu options, black or white; no looping option; and there are no options for adding a logo or title graphic. For standard-definition authoring, you might want to consider another application like iDVD, in which your project should not be compressed in FCP using the DVD option. Rather, export at high resolution and allow iDVD to do the encoding for you.

NOTE

Chapter Markers

FCP does not currently provide a mechanism for adding chapter markers to projects for export. This greatly limits its usefulness for disc delivery. Unfortunately, it also limits functionality for other delivery mechanisms, such as podcasting and even QuickTime, that allow chapters, letting the user jump between sections of the program. Chapter markers, however, can be added in Compressor or GarageBand.

YouTube, Vimeo, and Others

The next two groups in the **Share** menu are all account-based output options, whether it's **Podcast Producer**, which requires a server account; **Email**, which requires an email account; or accounts for YouTube, Vimeo, Facebook, or CNN iReport (see Figure 14.11).

FIGURE 14.11

Vimeo Share Option.

While the Email Share function produces pretty small files, as you might expect, the others are all set to automatically export a file at the original size. If you uncheck **Set size automatically**, you get the option to export to a different size. The options you get depend on the format you're sharing. On YouTube and Vimeo, you can upload at native resolution or, if the file is too large and beyond established upload limits, send a low-compression master, as these hosts will re-encode and optimize your material to their specifications.

You have very little choice in the compression options, either **Faster encode** or **Better quality (multi-pass)**. If you need better control over your exporting options, you really need to look to Compressor or some third-party encoding software.

EXPORT

There are four export functions. The primary one, which should be used for most export functions, is **Export Media**. This is the most important function because of capabilities built into it. It is important to understand how this function works. If your playback preferences are set to use original or optimized media, you can output a high-resolution master. If you are using proxy media, and your playback is set to proxy playback, you will output a proxy file, either in the ProRes Proxy codec, if you use Current Settings, or a compressed version of the proxy media. It is important that you make sure you are set to play back original or optimized media before you export. Also, if you have switched back and forth between proxy and original media, it is important that you check your project properties before exporting.

1. Use **Command-J** to open the Project Properties.
2. Click the wrench button in the lower right and check that the **Render Format** pop-up is set to **Apple ProRes 422**.

Export Media

The Export Media function allows you to export a high-resolution version of your project. When you have completed a project, I suggest that it should be standard procedure to use the **Export Media (Command-E)** function to produce a high-resolution, high-quality master that you can use to reproduce your project or recompress to other formats with minimal degradation of your image. The Export dialog (see Figure 14.12) allows you to output video and audio, video only, audio only, or roles, which we'll talk about in a moment. If you export a video file, in the **Video codec** pop-up, you get to choose from a number of available codecs (see Figure 14.13).

Which codec to use? Use **H.264** if you want a good-quality but compressed QuickTime video in the same size as the original. The file is probably too large for web delivery but might be suitable for a high-speed LAN. It's a smaller file than that shared to YouTube and Vimeo, but larger than that used by Media Browser and Apple

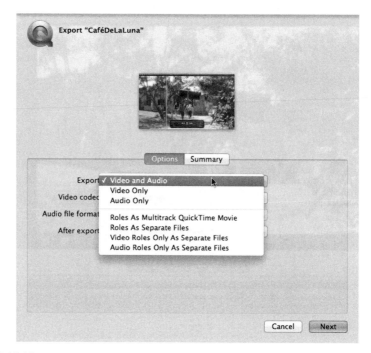

FIGURE 14.12

Export Dialog.

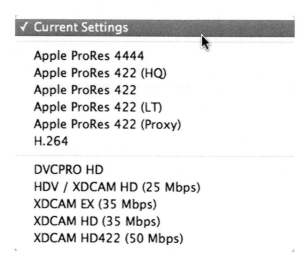

FIGURE 14.13

Available HD Codecs.

devices. It should not be used to make a project master. Use **Apple ProRes 422 (Proxy)** if you want to export a low-resolution proxy file, which is good for training videos that require heavily compressed media to fit onto a disc; otherwise, it's probably not something you're going to use.

In my view, **Apple ProRes 422 LT** (LT meaning Light) is an outstanding codec. I wish FCP would allow transcoding to this codec for Optimized Media because it is an excellent compromise of high quality with moderately high data rate. For most media from 24p formats or 720p or from AVCHD, this would be an excellent codec for importing. Unfortunately, it's not available for that. If you're going to continue editing on another platform, this may be a good export choice for you. For most purposes, the **Apple ProRes 422** codec, the standard current settings for render output, should be your choice for creating a high-resolution master file of your project. Anything edited in HD that's 1080p or smaller should be exported to this codec. If you are working in larger formats that are 2K or 4K in size, more than 2,000 or 4,000 pixels across, such as used in film formats or by the iPhone 4S, then you would be best served exporting to the **Apple ProRes 422 (HQ)** codec (HQ meaning High Quality). Be warned, though: These are very large, very high data rate files and require very fast drives to support them for proper playback. Time to look to those Thunderbolt drives! Finally, you have **Apple ProRes 4444**, which is used when you need to export a QuickTime movie that contains transparency. If you want to carry that transparency over to another application, then this is the codec of choice for you.

The other codec options below the line are for specialist proprietary formats used by Panasonic, Sony, and others.

When you want to export SD material, there is a different set of export options in addition to the ProRes codecs, including the **DV** and **DVC50** codecs. For creating an SD master of your project, I recommend using **DVC50**, though the file sizes are twice that of standard DV. It is a robust 4:2:2 codec that stands up very well to duplication and creates good-quality output when compressed.

NOTE

What Is 4:2:2?

The term 4:2:2 describes the way color is sampled in your image. The 4 is the luminance value, which is the most important component in reproducing the image. The 2s represent the Cb (blue channel) and the Cr (red channel). These are sampled at half the rate as the luminance channel. The green channel is not sampled at all and is calculated as a combination red, blue, and luma channel. This allows the image to be compressed by excluding some of the information. Other formats such as NTSC DV and DVCPRO25 use different compression schemes, such as 4:1:1, that are even more heavily compromised. This is one reason why these formats are very difficult to key properly. Other formats, such as PAL DV, MPEG-2, HDV, and XDCAM EX use a 4:2:0 scheme that is very difficult to work with in postproduction. Camera material in these formats is often immediately transcoded into ProRes 422 formats for editing.

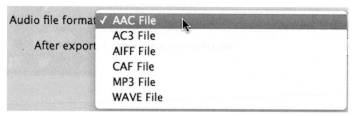

FIGURE 14.14

Audio Codecs.

> **NOTE**
>
> Interlacing
>
> If you shoot interlaced video, when you export to the Share options that compress the video, the material is deinterlaced. But when you export and want to make a full-size file without interlacing, you have to set that up manually. Go to the project's properties using **Command-J** and click the Wrench button on the lower left of the panel. In the dialog that opens, change the format to a *p* for progressive format to output a deinterlaced version.

> **TIP**
>
> Anamorphic Material
>
> iDVD still does not support the use of anamorphic material from DV, only native widescreen HDV. To get around this problem, you need to use software like Anamorphicizer or the excellent myDVDedit. You can also use the QuickTime 7 Pro player to set the frame size to a 16:9 aspect like 853×480 for DV NTSC or 1024×576 for DV PAL.

If you are exporting audio only, either as a single file or as roles, you get different audio export codec options. For an audio-only export, you get the codecs in Figure 14.14. They allow for compressed versions of your audio output: AAC and MP3, which are suitable for iPods; AC3, which is used for DVD production; CAF, for 5.1 Surround audio; and AIFF and WAV, for uncompressed audio, which should always be your first choice if you're making an audio master or exporting your audio for use in another production application.

Roles

Exporting roles allows you to create what are called *stems*, separate tracks of audio based on the content—for instance, separate stems for dialogue, effects, and music for professional audio mixing. Once you've created and assigned roles in either the Event Browser or in the Timeline, these assigned roles carry over to the exporting media function. You can export **Roles As Multitrack QuickTime Movie**, which will create a QuickTime file using any of the available ProRes and proprietary codecs, with each role laid on a separate track. You can also export **Roles As Separate Files**, which will give you QuickTime files, one for video and one for titles, as well as separate AIFF or WAV files for the audio stems. You should be aware, however, that if you're exporting separate tracks for video and titles, you must export using the ProRes 4444 codec; otherwise, you will lose title transparency. This is also true, of course, of the **Video Roles Only As Separate Files** option. If you want a single video track with combined titles and separate audio stems, you will need to do this in two passes: a video-only export in the codec of your choice and then a separate export of your stems for audio only using **Audio Roles Only As Separate Files**. Perhaps in later versions of the application you'll be able to specify which roles you want to export and which codecs for each.

NOTE

Export AAF/OMF

The Export AAF/OMF file formats are used to talk between video editing workstations and professional audio mixing and sweetening workstations, such as ProTools. While you cannot export to these formats directly from FCP, Automatic Duck (www.automaticduck.com/products/) has a free application that allows you to drag an FCP project from the Project Library directly to its application to create files that can be used in ProTools. The application, which was originally priced at $495, comes with excellent documentation and is now entirely free.

Still Image Export

There are two types of still image exports: either **Save Current Frame** or **Export Image Sequence**. To export a still image, you must be in the Timeline. You cannot export a current frame from the Project Library. In the Current Frame export dialog (see Figure 14.15), notice the checkbox for **Scale image to preserve aspect ratio**. This appears if you're exporting a format that does not use square pixels, such as standard-definition DV or some HD formats that record sizes that are 1440×1080 but display as 1920×1080. When you check the box with NTSC DV media, the exported file becomes 640×480 to preserve the aspect ratio, and the HD media becomes its display size of 1920×1080. If you want to keep the original size and aspect ratio, leave the checkbox open (unchecked).

You can export in a number of different still formats. Common ones include the default PNG, or TIFF, JPEG, or Photoshop, and more exotic ones include DPX, IFF, and OpenEXR.

TIP

De-interlace

You can also de-interlace a clip in FCP, by going to the clip's Info panel in the Inspector. In the lower left, change the pop-up to **Settings View** and set the **Field Dominance Overwrite** to **Progressive**. Don't forget to switch it back to the correct field dominance after you've exported the still image.

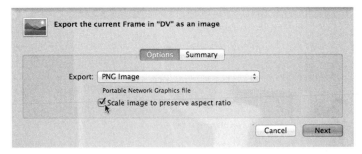

FIGURE 14.15

Still Export Dialog.

If you're going to export stills that come from video for web or print work, especially video with a lot of motion in it, you'll probably want to deinterlace it. You should do this in a graphics application such as Photoshop after you've exported. In Photoshop, it's in the **Filters** menu under **Video>Deinterlace**. In the Photoshop Deinterlace filter, you have an interpolation option, which works very well. There are some tools, such as PhotoZoom Pro from BenVista (www.benvista.com), that can enhance your still exports to hold up better in larger print sizes.

Image sequences are useful for rotoscoping, frame-by-frame painting on the video image, and other animation work, and they provide high-quality output without loss. The frame rate used for this will be the same frame rate as your project properties. Make sure you first create a folder in which to put your image sequence because this can easily generate a huge number of files.

HTTP Live Streaming

HTTP Live Streaming is an Apple implemented streaming format that allows simultaneous Internet streaming of multiple data rates. The device adapts to select the best available stream based on the available bandwidth. In the export module (see Figure 14.16), you get to select the various sizes you wish to export as well as the segment sizes.

Compressor

The **Send to Compressor** function, while it sounds like a good idea, gives you no significant benefit, while it adds substantially more time to the processing and encoding of your media. It is about 30% to 50% slower than exporting a master file and taking the master file to Compressor, and that includes the time to export the master file. It works like this: You select the function; Compressor 4 launches; and you get to choose whatever options, either Apple presets or custom presets, you've created. You set off the Compression. The problem is Compressor is continuously calling on FCP for frame information, which seems to be the bottleneck. I prefer to create my master and take that to the encoding application.

On the other hand, the **Export Using Compressor Settings** function works well (see Figure 14.17). This gives you access to all the QuickTime options available in

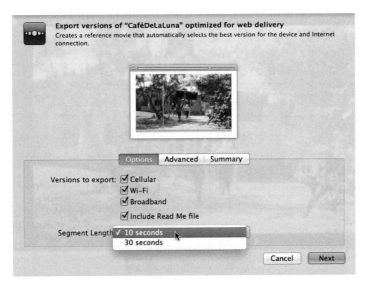

FIGURE 14.16

HTTP Live Streaming Export.

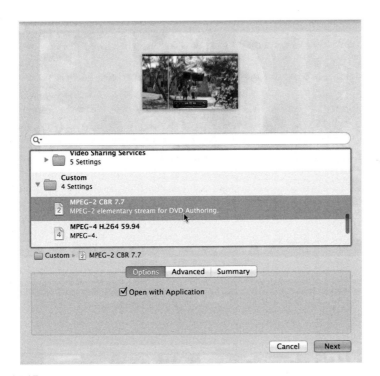

FIGURE 14.17

Export Using Compressor Settings.

earlier versions of FCP and FCE and even iMovie. Here, you can choose whatever format you want and customize it as needed. You simply need to create the custom preset in Compressor first. This does not launch Compressor at all but simply uses its setting presets. This is an excellent way to make custom export formats and is much, much faster than using the Send to Compressor function. However, Apple does require you to spend $50 to purchase what is essentially FCP's QuickTime export module, but if you need to make custom formats and custom sizes using cropping, then it's well worth the cost.

ARCHIVING

Now that you've finished your project, you should think about archiving your material. Before doing so, consider exporting a full-resolution native QuickTime file of your edited program.

The original memory cards, disks, or tapes on which you shot the project should have been correctly backed up already and should be your primary archive.

To archive the project, you need the project file and the associated events. One good way to do this is to duplicate your project, saving it onto a backup drive you use to archive your projects. You should duplicate it together with the associated events, either the full events or just the used clips, whichever you prefer. There is no need to include the render files. Once you have that archive made, a further backup can be made of just the event with its current version, but without any media, together with the project and its current version, but without any of its render or cache files. These should be relatively small files and can be burned to a disc for storage. This is a last-resort fallback. Should your archive drive fail, you can mount the secured event and project and then, inside FCP, use the **Import>Reimport from Camera/Archive** function to bring the video back from the camera archives into the event. Once that's done, the project will reconnect automatically to the material in the event, and you will have brought back all your work.

It's important that in addition to archiving your camera originals, you also properly secure backups of your graphics and audio files, including your FCP audio recordings, in separate locations outside the events. Digital media redundancy is the future of production. Whether it's on disc, LTO, hard drives, SSDs, or servers in the cloud, redundancy in your digital life is essential.

TIP

SuperDuper!

A handy way to archive your projects and media is to use Shirt Pocket's (www.shirt-pocket.com/) SuperDuper!, which allows you to schedule and back up your media drive items as well as selectively choose which folders to back up. This means you can exclude unnecessary and ever-changing render folders.

SUMMARY

We've now gone through the whole cycle of work in Final Cut Pro X, including setting up your computer; working with the interface; importing raw material, either analog or digital; editing; and adding transitions, titling, special effects, motion effects, color correction, and compositing. Now, finally, we have output our finished work and shared and exported it for everyone to see and enjoy. It's been a long road, but I hope you found it an exciting, interesting, and rewarding one. Good luck with all your future video projects and with your work in Final Cut Pro!

Index

Page numbers followed by *f* indicates a figure and *t* indicates a table.